本专著的出版得到以下基金项目资助:
国家自然科学基金面上项目，编号: 41771406，41471300

遥感分析与微气象模拟:

以埃边境干旱生态区地表热异常研究

覃志豪 著

Remote Sensing and Micrometeorological Modeling

——A study of thermal anomaly in an arid region across the Israel-Egypt border

Zhihao Qin

China Agricultural Science-Technology Press

图书在版编目（CIP）数据

遥感分析与微气象模拟 / 覃志豪著 . — 北京：中国农业科学技术出版社，2020.8
ISBN 978-7-5116-4896-9

Ⅰ . ①遥… Ⅱ . ①覃… Ⅲ . ①遥感技术 – 应用 – 微气象学 Ⅳ . ① P4

中国版本图书馆 CIP 数据核字（2020）第 137010 号

责任编辑　于建慧
责任校对　李向荣

出 版 者　中国农业科学技术出版社
　　　　　北京市中关村南大街 12 号　邮编：100081
电　　话　（010）82109708（编辑室）（010）82109702（发行部）
　　　　　（010）82109709（读者服务部）
传　　真　（010）82106650
网　　址　http://www.castp.cn
经 销 者　各地新华书店
印 刷 者　北京建宏印刷有限公司
开　　本　710mm × 1 000mm　1/16
印　　张　17
字　　数　323 千字
版　　次　2020 年 8 月第 1 版　2020 年 8 月第 1 次印刷
定　　价　100.00 元

Content

Preface

The frequent and long-term conflicts of Israel and Palestinian Harmas in Gaza Strip, a small area on the eastern shore of the Mediterranean Sea, has made the small Strip being one of the most known regions in the world due to its frequent appearance in the top news over the world. Just beside Gaza Strip there is a sand dune region stretching from the Israeli Negev into Egyptian Sinai. Locating in eastern coast of the Mediterranean Sea, the sand dune region in the Negev-Sinai peninsula is a typical arid land ecosystem, where two opposite directions of ecosystem evolution in the world can be observed over the sand dune region across the political border between Israel and Egypt. The Egyptian side is a typical example of desertification as a result of intensive grazing activities in the sand dune region by the nomadic Bedouin people. The sand dune region on the Egyptian Sinai side is mainly featured with bare surface with very little vegetation cover. Consequently the sand dunes are characterized as migratory under the wind blowing. As a contrast, the region in the Israeli Negev is under conservation without many anthropogenic activities, leading to the sand dune ecosystem undergoing its natural evolution with many shrubs, annuals and biogenic crust covering on the sand surface.

A very strange thermal anomaly was observed in satellite images in the sand dune region across the Israel-Egypt border. The Israeli side with more vegetation cover has obviously much higher surface temperature during daytime than the Egyptian side with much more bare sand surface. A series of puzzles are waiting to be revealed in answer to this strange phenomenon of thermal anomaly over the ground surface of the sand dune region.

Why does the thermal anomaly occur in the region? Is it a natural phenomenon or a man-made one? How does the anomaly change in the region? How does it vary among different seasons of the year? Are there any regularity in the spatial distribution and seasonal variation? What is the mechanism behind the occurrence of the phenomenon? The objective of this study is to uncover the mystery behind the thermal anomaly through a thorough examination from remote sensing analysis, ground truth measurement, numerical simulation and micro-meteorological modeling.

The book is mainly composed of my studies of the above questions as a PhD dissertation submitted to Ben Gurion University of the Negev (BGU), Israel. In August 1996, I came to BGU to pursue my PhD study in the Department of Geography and Environmental Development. In February 2000, I completed the study and submitted the dissertation to the university to apply for the degree of Doctor of Philosophy. On the 12th day of December, 2001 I was awarded the PhD degree by the university after critique reviewing, revision and evaluation for more than one year.

My study in BGU was a happy and memorial experience in my life. During the study I stayed at the Sede Boker Campus of BGU to carry on my researches in my supervisor Dr. Arnon Karnieli's Laboratory of Remote Sensing affiliated to J. Blaustein Institute for Desert Research (BIDR). The Sede Boker is a small village locating in the center of the southern Israeli Negev desert. However, the BIDR is a famous research institution on desert and arid environment in the world. The small village of Sede Boker is also famous and world-wise well-known due to not only the BIDR, but also due to the David Ben Gurion, who led the Jewish Zionism to the foundation of the State of Israel in 1948 and acted as the 1st Prime Minister of the State of Israel hence appreciated as the father of the State of Israel. The Gravesite of David Ben Gurion and his wife Pola Ben Gurion was located on the cliff of the Nahal Zin canyon just beside the Campus. Not far from the Campus is the communist village Sede Boker Kibbutz, which Ben Gurion joined after he retired from the State leader of Israel.

In the late 1990s, there were about 100 students, postdoc and visiting scholars from all over the world to do their research at BIDR. Among them Chinese students became the largest group. And we had about 30 Chinese students and their family members in the Sede Boker campus. The Chinese student group usually got together during the weekends and the Chinese festivals such as Moon Festival and Spring Festival. The 2nd year after I came to Israel, my wife Wenjuan Li and my son Max Sili Qin came from China in 1997 to join with me at Sede Boker. We had a very happy schooling life during my PhD study at the Campus. We had several close friends and their families at Sede Boker. Among them were Dr. Wenguan Zhao and his wife Zhenghui Li and their son Ze Zhao, Dr Genlin Jiao and his wife Xiaoping Fu and their son Hai Jiao, Dr. Fenchun Zhang and his wife Qian Shi and their daughter Feifei Zhang.

It can be said that my study at Sede Boker was ended with fruitful results. During this study, I published several articles that fixed my reputation in the academic realm of remote sensing. In order to compute land surface temperature (LST) of remote sensing data of Landsat TM used for the study, I developed a mono-window algorithm for LST retrieval from the TM data and published it in the International Journal of Remote Sensing. This algorithm has latter became well-known in the thermal infrared remote sensing circle due to fact that it was the 1st LST algorithm in the world for only one thermal infrared band data. I gave a complete derivation of the LST algorithm in the article, which was also part of my PhD dissertation presented in the book.

The way for me to reach my goal of willing for a PhD study was a little bit uneven. I finished my high school study in 1979 in Binyang Middle School, which was one of the top high schools in Guangxi Autonomous Region, South China, and was luckily passed the 1979 Chinese High Education Examination to enter Sun Yat-Sen University in Guangzhou of South China to study for BSc degree at the Department of Geography majoring in economic geography. After graduated from Sun Yat Sen University I came to Beijing to be a research scientist in the Institute of Agricultural Natural Resources and Regional Planning, Chinese Academy of Agricultural Sciences (CAAS) in the fall of 1983. I did various researches through projects in agricultural development. Though my researches were going smoothly and my position was uplifted from research assistant to associate research professor in early 1990s, but I was still keeping a dream of pursuing PhD study and continuing to look for opportunity of the study. In 1994 I was given a chance to come to Israel to participate in a 3-months short-term international training course in the Development Study Center in Rehovot, Israel, where I wrote to my supervisor Dr. Arnon Karnieli to express my willing of coming to his laboratory for PhD study. After reviewing my application, Arnon believed that I might have the potential of PhD study and started to help me for applying the required funding for the study.

During the long way to stride for the study, the supports from my wife and son are of the first importance. Without their rich love and supports, I could not finish my goal of PhD study with a happy memory experience. Therefore, my first thank was given to my family, especially my wife Wenjuan Li for her spiritual support and practical assistance in life especially taking care of our small family.

When I did my researches in CAAS, I still had a motive to pursue PhD study. This motivation was establelished from the situation that I as a young people in the 1980s of China was going along with the mainstream of the country's opening policy to realize a modernized country. This ideology pushed me to set up my willing to enhance my knowledge through study so that I could contribute more to the modernization of the country. Another implicit motivation was the fact that I needed to stride for opportunities to improve our life through increase of income. Going abroad for the study was my first choice of willing since this would simultaneously help me to reach the above two goals especially expansion of knowledge.

It was my supervisor Dr. Arnon Karnieli who dig me out from the applicants and offered the opportunity of coming to BGU for the study. Thus my second thank was given to my supervisor Dr. Arnon Karnieli for his kindly offering me the opportunity that led to my dream of PhD study for years coming into a true action in 1996 and for his wisdom of guiding me in the correct direction of thermal infrared remote sensing for the PhD study. I also would like to thank my other two supervisors Dr. Pedro Berliner and Dr. Haim Tsoar for their excellent supervision of my PhD study and giving me helps in many aspects during my study.

The International Center for Desert Research, J. Blaustein Institute for Desert Research, Ben Gurion University of the Negev offered me the fellowship for my PhD and the tuition for the first year study. The Department of Geography and Environmental Development, Ben Gurion University of the Negev (BGU), offered me the tuition for the second and the third year study. These financial supports provide a firm guarantee for my successful study at the university. I would like to express my grateful to the Center and the Department.

The colleagues at the Remote Sensing Laboratory offered a lot of helps in my study of the PhD dissertation. Svetlana Gilerman, the technician, helped me to use the computer facility of the Laboratory for my study. Heike Schmid and Giorgio Dell

Olmo, my two classmates under the same supervisor Dr. Arnon Karnieli, helped me travelling to the Nizzana Research Site of the study region for ground truth measurements and doing image processing for my study. Specially, Giorgio Dell Olmo discussed several academic issues with me for mutual understanding. Prof. Anatoly Getiosan and Dr. Dan Blurmburg, the faculty of the Remote Sensing Laboratory, offered me instructions through their excellent courses separately in remote sensing of environments and image process for remote sensing. For all of them, I sincerely thank.

I also thank the following persons for their various helps. Dr. Simon Berkowics, the administrator of The Arid Econsystems Research Center, Inst. of Earth Sciences, The Hebrew University of Jerusalem, offered me the permission to conduct ground truth measurements in the Nizzana Research Site and provided me his rainfall data of the region. Dr. Axel Allaier, visiting scientist at Department of Geography, BGU provided my two-month wind speed data and Dr. Thomas Litmann, Institute of Geography, Martin Luther University, Germany one-month air temperature, air relative humidity, wind speed and soil temperature data of the region for running my simulation model. Prof. Eyal Ben-Dor and Mr. Noam, Dept. of Geography, Tel Aviv University helped me to operate atmospheric simulation model MODTRAN. Prof. Nizzim Ben Yosef gave me some suggestions about my study, especially the atmospheric simulation for LST retrieval from AVHRR data. Prof. Natan Pokeika, Ben Gurion University of the Negev, allowed me to use his LOWTRAN 7 program for atmospheric simulation.

I had very good experience and life at Sede Boker. This was impossible without the friendly atmosphere of the Chinese student group at the Campus. I would like to thank our close friends Wenguan Zhao, Genlin Jiao, Fengchun Zhang, Xiaoping Fu, Qian Shi, Zhenghui Li, Zhenying Huang, Guoxiong Chen, Sheng Zhang, Qi Lu, Bo Wu, and many others for their sharing the Sede Boker life and assistances for the life.

My going to Israel for the study was also supported by my old colleagues and directors of the Institute of Agricultural Natural Resources and Regional Planning, CAAS. Herewith I also would like to express my grateful to them, especially the Institute's directors Yingzhong Li, Meiying Yang, and Huanjun Tang, and the colleagues Zhongyu Zhu, Yinjun Chen, Xiaoqing Huang, Changbin Yin, Jiangou Zhu, Shendi Su, Shunyi Wei, Qiyou Lou, Xuying Zhou, Tao Tao, Wenlai Jiang, Yuyun Bi, and many others. I wish all of my old colleagues healthy, have a good life, healthy body and smile face every day when the sun rises!

Publication of this book is supported by my current colleagues especially Dr. Zhaoliang Li, Dr. Maofang Gao, Dr. Sibo Duan, Dr. Pei Leng, Dr. Kebiao Mao, Dr. Xiaojing Han, and my students, especially Wenhui Du, Shifeng Li and Bilawal Abassi. To them, I would like to express my sincere thanks and hope them going smoothly in both directions to work and life.

Finally, the study may have many weaknesses and inadequacies due to my limitation of knowledge and data available. I do hope the readers to point out and make suggestions so that I can sharp it for a common understanding to the relevant issues under study.

Qin, Zhihao

August 20, 2019

Abstract

In the sand dune region across the Israel-Egypt border, an interesting phenomenon of thermal anomaly was observed on remote sensing imagery: the Israeli side with much more vegetation has a higher surface temperature than the Egyptian side, where bare soil and sand surfaces prevail. This study intends to examine the phenomenon by combining remote sensing analysis with micrometeorological modeling. The objectives are to analyze the seasonal and spatial change of land surface temperature (LST) on both sides and to model the surface temperature change and heat flux variation from the viewpoint of surface energy balance. The key factors and the mechanism leading the occurrence of thermal anomaly can be revealed through this examination.

Due to the availability, the remote sensing data of NOAA-AVHRR and Landsat TM has been used for the study. Retrieving LST from AVHRR data is mainly through a split window algorithm. Several algorithms had been proposed in literature for the retrieval. However, the application of the existing algorithms to the region faces many difficulties because they involve not only the most important parameters (atmospheric transmittance and ground emissivity) but also some parameters not easy to estimate due to the lack of proper *in situ* atmospheric profile data.

A new split window algorithm only containing the two most important parameters has been developed in the study for retrieval of LST from AVHRR data of the arid environment. A complete derivation of the new algorithm has been presented with details, which includes the establelishment of thermal radiance transfer equations, linearization of Planck's radiance function, derivation process of the algorithm, determination of atmospheric transmittance and sensitivity analysis of the algorithm. Results from sensitivity analysis that the new algorithm can provide quite accuracy of LST estimation. For an error of 0.05 in atmospheric transmittance and 0.01 in ground emissivity, the error of LST estimation by this algorithm is about 0.4°C and 0.7°C, with a combination error of about 1.1°C, which is lower than the generally expected 1.5°C.

Using the new algorithm, the available AVHRR data between 1995 and 1998 had been processed for LST retrieval. Generally, the images with clear sky and good viewing angle ($\geqslant 60°$) were selected for the processing. The results indicate that LST of the region is much higher in summer than in winter. In summer, LST of the region may reach high up to 50-56°C while in winter it is about 28-35°C in early afternoon when the satellite passed to acquire the thermal data of the region. The nighttime LST of the region is about 20-25°C in summer and 10-15°C in winter. An obvious LST contrast can be seen on both sides during the day in summer. The Israeli side usually is about 2-3.5°C hotter than the Egyptian side. In winter the LST difference is weak (<1°C). The obvious LST contrast on both sides disappears during the night in summer because LST on the Israeli side is not always higher. Nighttime LST on the Egyptian side is slightly higher in winter.

Landsat TM has a thermal channel with a spatial resolution of 120 m×120 m, which is very suitable for analysis of spatial LST variation. Up to present the algorithm for retrieval of LST from the only one thermal band of Landsat TM data has not been published. Considered the characteristics of atmospheric profiles, a mono-window algorithm has been developed in the study for LST retrieval from Landsat TM data. Except the atmospheric transmittance and ground emissivity, this algorithm includes another important parameter: the effective mean atmospheric temperature, which can be easily estimated with the method proposed in the study. The derivation of the algorithm and determination of its parameters has been also presented with details. The sensitivity analysis indicates that the possible error of ground emissivity has an insignificant impact on the probability LST estimation error, which is sensitive to the possible error of transmittance and mean atmospheric temperature. However, average LST estimation error with this algorithm is generally within 1.5°C, which can meet the accuracy requirement of the study.

The mono-window algorithm has been applied to retrieve LST from available Landsat TM data of the region. The result indicates that the spatial variation of LST is very obvious on both sides, though the images were acquired around 9:30 in the morning when the surface is not very hot. The border can be clearly identified in the LST images. The hottest place locates in the Israeli side and the coolest in the Egyptian side, with an average difference of 2-3°C.

In order to validate the satellite-observed LST change on both sides, the ground truth measurements have been conducted in various seasons during 1997-1998 at the Nizzana Research Sites, which is adjacent to the border (see Figure 5.4). Soil samples of sand and biogenic crust, the two most important surface patterns of the region, were taken from the field for experiments to determine their emissivity. The result indicates that biogenic crust has an emissivity of about 0.97 and sand 0.95 in dry condition. At saturation, these two surfaces tend to have an identical emissivity of about 0.985, which is slightly lower than the generally accepted water emissivity 0.99.

A hand-hold IR thermometer was used to measure the surface temperature in the field through sampling on the four surface patterns: biogenic crust, sand, playa and vegetation (shrub). The results from these measurements confirm what was observed on remote sensing imagery. In the dry summer at about noon, the difference of surface temperature between biogenic crust and sand is high up to above 3°C. The difference is also obvious in the morning and afternoon. In early and late dry seasons, the difference is still high up to above 2.5°C at about noon. The difference disappears only in the several days after heavy rain when the ground soil is very wet.

The observed LST anomaly on both sides is in fact the direct result of their structural difference of surface composition. Up to present the surface composition of the region has not been quantitatively reported though the obvious difference, especially vegetation cover on both sides has been mentioned in several studies. Three methods have been employed for the measurement of surface composition on both sides: field observation, measuring on the aerial photograph with very high spatial resolution and determining the vegetation rate on Landsat TM image. Based on these measurements, I estimate that vegetation cover rate is about 17.5% on the Israeli side and 4.5% on the Egyptian side. Biogenic crust covers about 72% of the Israel side and 12% of the Egyptian side. Sand accounts for 7% on the Israeli side and 80% on the Egyptian side. Finally, playa covers about 3.5% on both sides.

According to the relationship between surface composition and temperature change, the average LST change and its difference on both sides has been estimated. The result indicates that the Israeli side has steadily higher LST under moderate to high temperature level (above 35°C). When the temperature is low such as several days after heavy rain in the wet season, the Israeli side will be cooler. The LST difference on both sides reduces to minimal when temperature level is within 17-27°C. This result explains why an obvious LST contrast is observed during the day and why disappears during the night in summer. It also answers the question why the Egyptian side has slightly higher LST in wet conditions.

In order to understand the inner functioning mechanism leading to the LST difference and the key factors controlling the anomaly, a simulation model has been establelished in the study from the viewpoint of surface energy balance. The purpose is to compare the surface temperature change and heat flux variation on the two most important surface patterns (biogenic crust and sand). The model couples soil temperature change with soil moisture movement to solve soil heat flux and latent heat flux. Thus, it includes two differential equations stating the two processes. The numerical solutions to the model and the differential equations have been developed in the study with detailed presentation of their derivations. This mainly includes the application of Crank-Nicholson implicit approximation method to simultaneously solve soil temperature and moisture from the differential equations as well as Newton-Raphson approximation to solve LST and heat fluxes from the model.

In order to run the model, several measurements have been conducted in preparation of the required data. These indude the global radiation measurement, surface albedo estimation, soil water content determination, air temperature, air relative humidity and wind speed measurements. The data prepared for the simulation was taken from the measurements in July 1998 to represent the general situation of the hot dry season of the region.

Results from the simulation indicate that biogenic crust does have an obviously higher surface temperature at about noon. The surface temperature difference between biogenic crust and sand is about 2-3°C from early morning to late afternoon. During the night, the two surfaces have very similar surface temperature change. This simulation is in accordance with the LST variation observed in the remote sensing images of the region.

According to the simulation, biogenic crust has much higher net radiation than sand during the day. This is mainly due to its lower surface albedo. Above 85% of net radiation dissipates as sensible heat flux into the air. Soil heat flux only accounts for 10%-15% of the net radiation. Due to little soil moisture available for evaporation, latent heat flux is very little (accounting for <1% of net radiation). The amount of sensible heat is positively proportional to surface-air temperature difference and negatively proportional to wind speed. Thus, the greater the sensible heat flux, the higher surface temperature under the same air temperature and wind speed. Even though biogenic crust has a greater latent heat flux than sand, the amount of evaporation is too small to cool down the surface. The simulation also provides the way to understand some micrometeorological phenomena and soil hydrological properties of the region such as dew formation during the night, soil moisture content difference between the two surfaces, surface moisture and humidity vibration with time, and so on.

Simulation with the combination of parameters of the two surfaces indicates that surface albedo is the key factors leading to the LST difference between the two surfaces, which is followed by the sub-surface soil properties (especially soil moisture content) as a whole. Albedo accounts for about 87.83% of the LST difference between the two surfaces and sub-surface soil properties as a whole explains about 10.27%. Though surface emissivity has significant difference on the two surfaces, hence on both sides, it can only account for about 1.90% of the LST difference between the two surfaces.

The mechanism of heating process difference on the two surfaces can be extended to understand the LST anomaly on both sides. During summer, the region is very dry and vegetation is in dormancy. The leaves of most shrubs reduce to a minimum and more than two third of the canopy is dry or dead. The dead canopy is not only out of functioning transpiration but also absorbs more incident radiation due to its darker tone hence lower albedo. Even though the Israeli side has more vegetation, the evapotranspiration from both ground and vegetation is still very small in comparison with net radiation. Total evapotranspiration from 12:00 to15:00 on the Israeli side is estimated to be only about 7% of its net radiation. This small latent heat flux has little effect on the surface energy balance process in the arid environment. LST change in the desert region is mainly controlled by the amount of incident solar energy absorbed by the ground as soil heat. Therefore, the anomalous LST change and thermal variation on both sides of the region still can be explained as the direct result of their obvious albedo difference. This albedo difference is mainly caused by the sharp contrast of surface composition on both sides. As mentioned above, the Israeli side has much more biogenic crust while the Egyptian side is mainly dominant with bare sand. Considered this sharp contrast, the LST anomaly observed on remote sensing images can be understood as the result of spectral functioning differences of biogenic crust and sand on both sides of the region.

Abbreviation and Notation

1. Abbreviation

AVHRR	Advanced Very High Resolution Radiometer
DN	Digital number
IR	Infrared
LST	Land surface temperature
KST	Kinetic surface temperature
MSS	Multi-Spectral Scanner
NASA	National Aeronautics and Space Administration
NIR	Near infrared
NDVI	Normalized difference of vegetation index
NOAA	National Oceanic and Atmospheric Administration
RMS	Root mean square
RST	Radiant surface temperature
TM	Thematic mapper
ZAV	Zenith angle of viewing

2. Notation in remote sensing analysis

a_0, a_1, a_2	Coefficients of split window algorithm in Eq. (2.19)
a_i, a_4, a_5, a_6	Parameters for channel i, 4, 5 of AVHRR and band 6 of TM
A	Coefficient of split window algorithm
A_0, A_1, A_2	Coefficients of split window algorithm in Eq. (3.29)
b_i, b_4, b_5, b_6	Parameters for channel i, 4, 5 of AVHRR and band 6 of TM
B	Coefficient of split window algorithm, K
$B_i(T_i)$	Radiance received by the sensor, W m^{-2} sr^{-1} μm^{-1}
$B_i(T_s)$	Radiance emitted by ground
$B_i(T_z)$, $B_i(T_a)$	Atmospheric radiance at T_z and T_a
$B_\lambda(T)$	Planck radiance function of the blackbody, W m^{-2} sr^{-1} μm^{-1}

c	Parameter in Eq. (2.19c)
c_1	The first spectral constants in Eq.(2.1), c_1=5.955215×10^{-17} Wm^2
c_2	The second spectral constants in Eq.(2.1), c_2=1.43876869×10^{-2} m K
C_i, D_i	Parameters in deriving split window algorithm in Eq. (3.10)
E_1, E_2, E_3	Parameters for simplifying coefficient B
i	Thermal channel of AVHRR, i=4 or 5
$I_i^{\downarrow}$, $I_i^{\uparrow}$	Down-welling and upwelling atmospheric radiance
k_i	Atmospheric parameter in Eq. (2.13b)
K_1	Pre-launch calibration constants of TM6, K_1= 60.776 mWcm^{-2}sr^{-1}μm^{-1}
K_2	Pre-launch calibration constants of TM6, K_2=1260.56K.
L_i	Parameter of Taylor's expansion in Eq. (3.12)
$L_{\max(\lambda)}$	Maximum radiance of TM6, $L_{\max(\lambda)}$=1.56 mW cm^{-2}sr^{-1}μm^{-1}
$L_{\min(\lambda)}$	Minimum radiance of TM6, $L_{\min(\lambda)}$=0.1238 mW cm^{-2}sr^{-1}μm^{-1}
$L_{(\lambda)}$	Radiance converted from the DN value of TM6, mW cm^{-2}sr^{-1}μm^{-1}
M, P	Coefficients of split window algorithm in Eq. (2.17)
n_i	Parameter in Eq. (2.7) and (3.14)
Q_{dn}	DN value of TM6
Q_{max}	Maximum DN value of TM6, Q_{max}=255
R_v	Vegetation cover rate
$R_t(z)$	Attenuation rate of atmospheric temperature at altitude z, K
$R_w(z)$	Ratio of water vapor content at altitude z to the total, K
$TISI_n$, $TISI_d$	Temperature independent spectral index for night and day
T	Temperature, K
T_a	Effective mean atmospheric temperature, K
T_i, T_4, T_5	Brightness temperature of channels i, 4 and 5 in AVHRR, K
T_r	Radiant temperature, K
T_s	Land surface temperature, K
T_{bs}, T_v	Surface temperature of bare soil and vegetation canopy, K
T_z	Atmospheric temperature at altitude z
T_6	Brightness temperature of Landsat TM6, K
w	Water vapor content in atmospheric profile, g/cm^2
$w(z)$	Water vapor content in atmospheric profile at altitude z, g/cm^2

W_i	Atmospheric parameter in Eq. (2.9)
Z, z	Altitude of the sensor and the layer between ground and sensor, km
$\Delta\tau_i(\theta)$	Transmittance correction by zenith angle of viewing
$\Delta\varepsilon$	Emissivity difference between channels 4 and 5 of AVHRR
ε_i, ε_4, ε_5	Emissivity of channels *i*, 4 and 5 in AVHRR
γ	parameter in Eq. (2.19)
λ	Wavelength, m
θ	Zenith angle of viewing, degree
θ'	Zenith angle of downward atmospheric radiance
δT_s	LST estimation error, °C
δx	Possible error of variable *x*
$\delta\tau_i$	Transmittance error of channels *i*
$\delta\varepsilon_i$	Emissivity error of channels *i*
$\tau_i(\theta)$	Total atmospheric transmittance at θ
$\tau_i(\theta, z, Z)$	Upwelling atmospheric transmittance
$\tau_i(\theta', z, 0)$	Down-welling atmospheric transmittance

3. Notation in micrometeorological modeling

a_i, b_i, c_i	Coefficients in equation (6.5.1)
A_i, B_i, C_i	Coefficients in equation (6.5.2)
c_a, c_m	Specific heat of air and clay, J kg^{-1} K^{-1}
c_q, c_w	Specific heat of quartz and water, J kg^{-1} K^{-1}
c_s	Specific heat of soil, J kg^{-1} K^{-1}
C_h	Change of soil water *vs* relative humidity, kg/m^3
C_s	Volumetric soil heat capacity, J m^{-3} K^{-1}
d_i, e_i, f_i	Coefficient in equation (6.5.1)
D_i, E_i, F_i	Coefficient in equation (6.5.2)
dT	Small change in temperature, K
D_v	Apparent vapor diffussivity, kg m^{-1} s^{-1} kPa^{-1}.
D_s, D_b	Soil particle and bulk density, kg/m^3
e	Soil vapor pressure, kPa
e_a, e_s	Air and surface vapor pressures, kPa

e_v	Saturated soil vapor pressure, kPa
g	Acceleration of gravity, g=9.8m/s^2
g_j	Parameter in equation (6.4.11)
G	Soil heat flux, W/m^2
G_j	Parameter in equation (6.4.14)
h	Soil relative humidity
h_i	Soil relative humidity at the layer i
H	Sensible heat flux, W/m^2
i, j	The ith and jth soil layer
k	Von Karman constant, k=0.4
K_a, K_b	Parameters in equation (6.3.12)
K_c	Soil hydraulic conductivity, kg s/m^3
K_d, K_e	Parameters in equation (6.3.15)
K_s	Soil thermal conductivity, W m^{-1} K^{-1}
L	Latent heat vaporization, L=2.543×10^6 J/kg
LE	Latent heat flux, W/m^2
LE_c	Energy driving soil water change, W/m^2
M_s	Soil mass, kg
n	Number of soil layers
P_a	Atmospheric pressure, kPa
P_s	Soil porosity
r_a	Air resistance coefficient to heat transfer, s/m
r_s	Surface resistance, s/m
R	Universal gas constant, R=461.52 J kg^{-1} K^{-1}
R_n	Net radiation, W/m^2
R_s	Global hemispheric radiation, W/m^2
s	Slope of saturated vapor pressure *vs.* temperature, kPa/K
T	Temperature, K
T_i,	Soil temperature at layer i, K
T_a, T_s	Air and surface temperature, K
TM	Thematic Mapper
u_z	Wind velocity, m/s

V_a, V_m	Volumetric fraction of air and clay in soil
V_q, V_w	Volumetric fraction of quartz and water in soil
V_s, V_p	Volume of soil materials and soil pore
z	Standard height, z=2m
z_0	Roughness length, m
Z_t	Parameter in equation (6.4.11)
ε_a, ε_s	Air and surface emissivities
φ_h, φ_m	Stableility correction parameters for heat and momentum
ρ	Surface albedo
ρ_a, ρ_m	Density of air and clay, kg/m^3
ρ_q, ρ_w	Density of quartz and water, kg/m^3
ρ_s	Soil density, kg/m^3
σ	Stefan-Boltzmann constant, $\sigma=5.67\times10^{-8}$ Wm^{-2}K^{-4}
ψ	Soil water potential, J/kg
$\partial h/\partial t$	Rate of soil relative humidity change, s^{-1}
∂t	Time interval, s
$\partial T/\partial t$	Rate of soil temperature change, K/s
∂z	Depth of soil layer, m
$\partial\theta/\partial t$	Rate of soil water change, kg m^{-3} s^{-1}

1 Introduction

Surrounded by Mediterranean Sea in the north-west and the Gulfs of Suez and Aqaba by Red Sea in south, the Israel-Sinai (Egypt) peninsula (Figure 1.1) is a unique and special region in the world. Historically, this region was the origination centre of the world's three main religions: Judaism, Islamism and Christianity and consequently it was appraised as the Holy Land. Contemporarily, the political and ideological conflicts between Jewish and Arab branded it as one of the hottest places in the world.

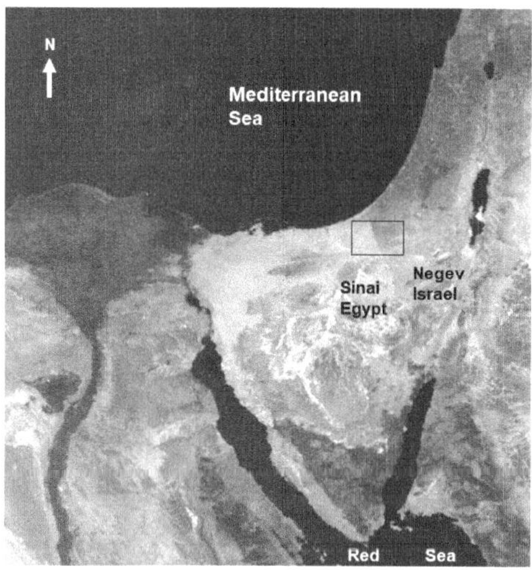

Figure 1.1 Remote sensing image of the Israel-Sinai peninsula, illustrating the spectral contrast on both sides of the Israel-Egypt border. This pseudo color image is composed from NOAA-AVHRR channels 1, 2 and 4 as RGB, acquired on July 18, 1998. The box inside shows the study region.

Standing geographically at the crossroads of Europe, Asia and Africa as well as locating geologically at the junction of the African continent and the Middle East plate, the region is not only interpreted as of geopolitical importance in global strategies but it is also characterized with obvious landscape and climate differences. Rich rainfall in the Mediterranean coastal region and northern Israel produces its unique climate pattern and dense vegetation cover features. In contrast, bare rocky mounts and valleys with extremely drought desert environmental ecosystem are the prevalent landscape in the central and southern Sinai. Three natural zones can be distinguished in the region: Mediterranean climate area, savannah, and desert. These unique features have attracted many scientific studies oriented to the region (Ritter, 1968; Ostendorff et al., 1976; Danin, 1983; Neev et al., 1987; Karnieli, 1995; Yair, 1997).

The northwestern part of the Israel-Egypt political border is crossed by linear sand dunes of the same lithological unit. The sand dunes were formed by the deposition of sand and dust transported from the North African desert by strong wind. In remote sensing images of visible channels such as Figure 1.2, this region is characterized by a sharp contrast between bright reflectance from the Egyptian side (Sinai) and dark one from the Israeli side (Negev) (Tsoar, 1996). The contrast was interpreted as being caused by different land surface cover structures under different land use policies (Otterman, 1974; 1977; 1981; Warren, 1984; Tsoar and Møller 1986). The Egyptian side has lower vegetation cover and more sand surface. On the contrast, the Israeli side is dominated with much more shrubs and inactive sand surface fixed by biogenic crust,

Figure 1.2 The Landsat TM image of the study region, composed of band 4, 3 and 2 as RGB. The spectral contrast is sharp on both sides of the Israel-Egypt border, which can be clearly identified with the highway along it. The sand dunes can also be seen in the image.

also named as the microbial crust (Kidron, 1997). The thin biogenic crust (<5 mm) mainly concentrates on the lower part of the longitudinal dunes in the western Negev, Israel and is very few on the Egyptian side due to its intensive anthropogenic activities (Tsoar, 1986). Similar spectral contrast also was observed along the United States and Mexico border (Bahre, 1978; Balling, 1989; Balling et al., 1998). The United States side was darker because of its more vegetation cover. The Mexican side is brighter because of its less vegetation cover resulted from severe overgrazing activities (Balling et al., 1998).

In regular conditions, one might think that since the Israeli side has more plants that contribute to regional evapotranspiration, it should be cooler than the Egyptian side. Such phenomenon has been observed across the border between USA and Mexico (Balling, 1989). However, an opposite thermal change was observed on the daytime thermal images of the Israel-Egypt border region. The Israeli side has a higher land surface temperature (Figure 1.3 and Figure 1.5). This phenomenon occurs not on a specific day, but in almost all the year except the days when the ground is very wet and cool. Considered more vegetation cover on Israeli sides, this phenomenon of temperature difference in the border region is really anomalous. The sharp contrast of land surface temperature (LST) on both sides cannot be seen on nighttime thermal images (Figure

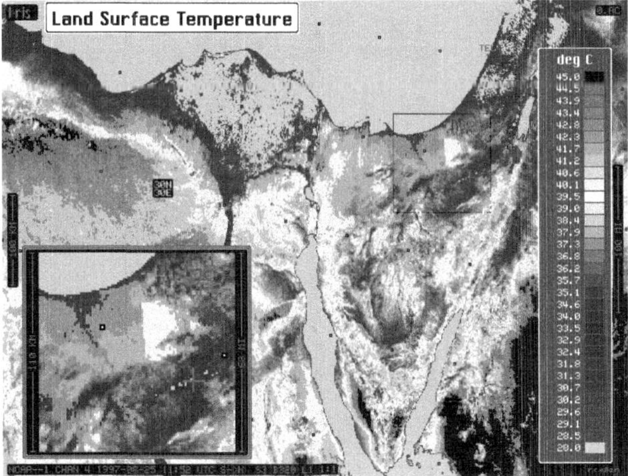

Figure 1.3 Daytime thermal images of the sand dunes across the Israel-Egypt border, computed from NOAA-AVHRR 14. An obvious temperature contrast can be clearly seen on both sides of the border.

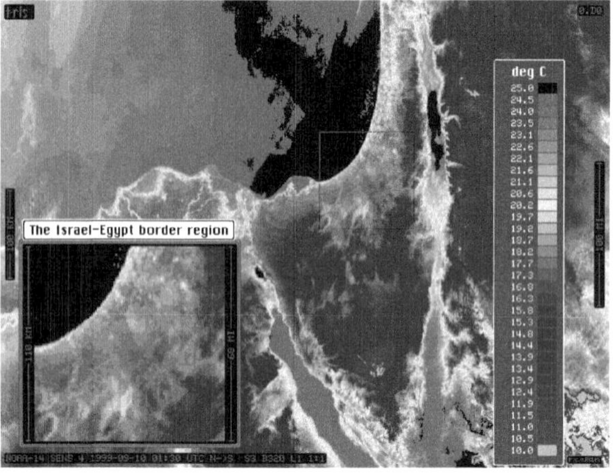

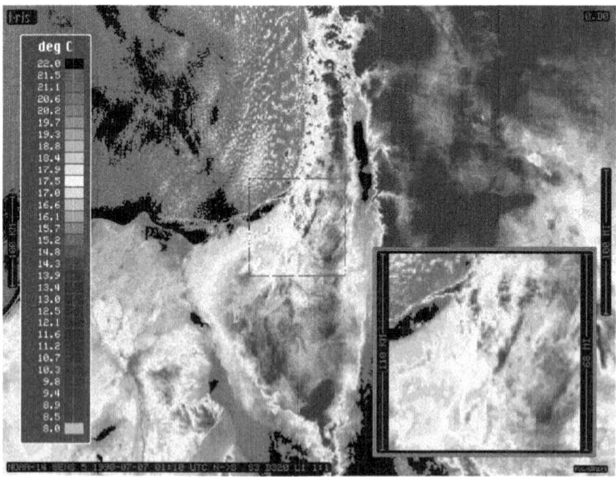

Figure 1.4 Nighttime thermal images of the sand dunes across the Israel-Egypt border, computed from NOAA-AVHRR 14 data. No obvious temperature difference can be seen on both sides of the border.

1.4)

What factors lead to the occurrence of this anomalous LST phenomenon during the day and disappearance during the night? How does the LST phenomenon perform in different seasons and different years? The biogenic crust is a common surface pattern in the Israeli Negev and many other arid regions such as the Sahara (Karnieli, 1997; Yair et al., 1997). Does the crust have any special function in controlling surface temperature change and surface energy balance of the arid region? These questions and related issues are the main questions that this study intends to examine and answer.

Figure 1.5 LST variation in the sand dunes across the Israel-Egypt border, retrieved from NOAA-AVHRR-14 data acquired at 12:08 GMT or 15:08 local time on July 18, 1998. The dimension of the number in the legend is Celsius degree. The boxes inside show the position of the subsets taken for analyzing the LST change on both sides. The subsets on both sides are in the same size.

1.1 The study region and the related studies on the issues

The study region as shown in Figure 1.1 covers an area of about 30 km^2×40 km^2 (30 km along the border and a width of about 20 km on each side). The geomorphological patterns of the region are mainly the linear sand dunes and interdune valleys stretching from the Egyptian side into the Israeli side. Fine sand with a size of 0.1-0.5mm (Tsoar, 1990; Gerson, 1985) is the principle material of soil constituents in the region, though silt and clay also accounts some percentages especially on the Israeli side where a thin biogenic crust covers most of its surface. Average annual rainfall of the region is 95mm (Kidron, 1997). Rain season concentrates on the period from November to April of next year.

In the Egyptian side (Sinai), the region is under intensive utilization by the Bedouin nomads, who use the sand surface for grazing goats, camels and other cattle as well as some activities of cropping. High plants (shrubs) have been subjected to asevere gathering of firewood (Tsaor,1986). In contrast to the free-use on the Egyptian side, the Israeli side (Negev) has been managed under strict conservation policies. The limited anthropogenic activities on the Israel side have led to the establelishment of vegetation (shrubs and annuals) and biogenic crust (composed of lichen, fungi and other microphytes especially chlorophyll-containing cyanobacteria) on the surface of the sand dune region (Karnieli, 1997).

The difference in land use between Israel and Egypt along the political border has a pronounced effect on remote sensing imagery. On the images of visible bands, A sharp spectral contrast can be seen between the two sides. This contrast has attracted many scientific efforts over the last thirty years (Otterman and Fraser, 1976; Tsoar and Møller, 1986; Danin, 1987 and 1991; Karnieli and Tsoar, 1995; Tsoar et al., 1995; Tsoar and Karnieli, 1996; Karnieli et al., 1996; Karnieli and Sarafis, 1996; Karnieli 1997). The relatively higher reflectance value of the remote sensing image on the Egyptian side is believed to be directly resulted from the severe anthropogenic impact of the Sinai Bedouin, including overgrazing and firewood gathering (Tsoar, 1986).

The traditional and popular explanation asserts that the spectral contrast between the Israeli and Egyptian sides is mainly due to the different patterns of anthropogenic impact on both sides. Otterman (1974) observed the contrast at the Sinai-Negev boundary in the first Landsat image of the area taken in 1972 and pointed out that semi-dormant desert fringe plants strongly reduced the albedo of sandy terrain. The contrast reported in that study was about the same in all spectral-bands of the Landsat Multi-Spectral Scanner (MSS), which included two visible (500-700nm) and two near infrared (NIR, 700-1100nm) bands. It was claimed that the plant cover on the vegetated side of the Negev did not exhibit characteristics common to green vegetation of non-arid regions. It was further stated that green plants would have produced a Sinai-Negev contrast ratio much higher in the visible bands and in particular in the MSS red band (600-700nm), where plant chlorophyll strongly absorbs solar radiation. The contrast was therefore interpreted as a product of dark plant debris and shadowing effects.

Karnieli and Tsoar (1995) suggested a different interpretation to the spectral contrast. In summary, three types of evidence have been combined to verify the hypothesis explaining the difference in spectral reflectance on the opposite sides of the border. (1) Vegetation cover in the region is less than 30% and therefore soil is believed to be the principal contributor to overall spectral reflectance of the region. (2) The spectral reflectance of the arid dune vegetation is much lower than that of the sand and biogenic crust. Consequently, the dune vegetation signal is masked by soil signal in terms of remote sensing. (3) The biogenic crust covering most of the Israeli side contributes a darker tone to remote sensing images. Anthropogenic activities on the Egyptian side that prevent the establelishment of the crust contribute to the brighter tone. Therefore, it is concluded that the well known contrast between Sinai and Negev in remote sensing imagery that has drawn the attention of many scientists is not a direct result of vegetation cover but caused by the almost complete cover of biogenic crust (Karnieli and Tsoar, 1995).

Tsoar and Karnieli (1996) studied the change of reflectance values on remote sensing. On the Israeli side of the sand dune region, the increment of biogenic crust generally comes together with an increase in vegetation cover. A decrease in the brightness values can be seen between 1984 and 1989 in the data of all Landsat MSS bands except the NIR band 7 (800-1100nm) which shows little change or no change with time. However, the Sinai bare sand shows the opposite trend of little to no change in all bands except band 7, which shows an increase in reflectance with time. This is probably due to the effect of further destruction of vegetation in Sinai after 1982. In the Negev part of the region, the sand dunes are low and stable with high percentage of biogenic crust in interdune. The crust is developed by microphytes from fines (silt and clay) on the surface. The region is covered by dense carpets of wilted annuals which have a reflectance that is lower than crusted area, but higher than the Artemisia shrub (Tsoar, 1996).

The contrast between the two sides of the Israel-Egypt border is also observed in the thermal range of the electromagnetic spectrum. Otterman and Tucker (1985) found that the vegetated exclosure or the sand with protruding shrubs has higher (by up to 3°C) radiant temperature than the surrounding terrain. They related this thermal contrast to severe overgrazing and other anthropogenic pressures on the Egyptian side. After analyzine this impact, they concluded that in an arid climate, under the semi-dormant conditions of vegetation, which prevail at all times except for the short desert blooming period after raining, evapotranspiration is very low and has apparently a negligible effect on the surface temperature. Thus, LST of desert region is controlled by the surface albedo (Otterman and Tucker, 1985). This explanation is reasonable, but is not enough because there is no direct evidence about the quantitative difference of albedo and evaporation on both sides. Furthermore, the mechanism leading the LST difference still remains unknown. Actually, the LST change of the ground surface is controlled by many factors such as sub-surface soil properties including soil thermal conductivity, soil water content and heat capacity. Micrometeorologically and soil-physically, the mechanism leading to LST change is rather complicated. Moreover, the phenomenon is observed in daytime of not only dry season, but also wet season when the plants (shrubs and annuals) are growing. Therefore, a study is necessary to explain why the anomalous phenomenon of LST change occurs in the region.

1.2 Objectives and hypothesis

The prime objective of the study is to examine, by combining remote sensing analysis with micrometeorological modeling, the LST change in the region, with particular concentration on analysis of its seasonal and spatial change and explanation to its difference on both sides.

The Remote sensing data used for the study mainly includes the thermal channels of Advanced Very High Resolution Radiometer (AVHRR) of the National Oceanic and Atmospheric Administration (NOAA) 14 satellite and Landsat Thematic Mapper (TM). AVHRR has two thermal channels (4 and 5) and TM has one thermal band (TM6) which can be used for the study. The objective of using the remote sensing data is to analyze the seasonal and spatial change of LST in the region. Though several algorithms have been proposed for LST estimation from remote sensing data (Qin and Karnieli, 1999), their application to the arid region still remains many problems. Thus, development of a methodology suitable for the region to retrieve LST from the remote sensing data is also one of the main objectives of the study. The methodology will then be used for retrieval of LST from both AVHRR and TM data for the analysis of its seasonal change and spatial variation.

In order to validate the remote sensing observation of LST change in the region, ground truth measurements on typical surface patterns are necessary to be conducted on the Israeli side. Sand and biogenic crust are the two most important surface patterns in the region. Therefore, the validation will be mainly oriented to the comparison of the LST difference on the two patterns, though other surface patterns (vegetation and playa) are also analyzed.

One of the hypotheses is that the LST observed by remote sensing can be directly ascribed to the difference of surface composition under the same lithological structure, which is exactly the case in the arid region. Thus, the measurements of surface composition structure and comparison of their difference on both sides can help us to understand its impact on the LST difference in terms of remote sensing.

On the other hand, LST in the soil-atmosphere system is determined by many micrometeorological and soil-physical factors through the dynamics of surface energy balance. In order to reveal the mechanism leading to the LST anomaly on both sides, it is necessary to establelish a micrometeorological model to simulate surface temperature change and heat flux variation of the region from the viewpoint of surface energy balance. Because it is impossible to access the Egyptian side for any field measurements and observations, all the research work has to be done on the Israeli side. Therefore, the simulation is mainly oriented to the two typical surface patterns.

Another hypothesis of the study is that the LST difference between sand and biogenic crust can be extended to represent the general case on both sides of the region. Application of the micrometeorological model to the two surface patterns for comparison of their difference in terms of surface energy balance can finally lead to the uncover of key factors controlling the sharp LST difference on both sides of the region.

More specifically, the hypotheses based on the micrometeorological modeling for the study can be summarized as follows.

- The difference of surface albedo, which is much higher on the Egyptian side, controls the absorption difference of incident global radiation on both sides. Under arid environment, evaporation is minimal due to extremely limited soil water content. Thus, the absorption difference of incident energy has significant contribution to the difference of LST on both sides.

- The difference of emissivity, which is expected to be higher on the Israeli side, has significant contribution to LST difference on both sides of the region. For the same temperature level, emissivity difference determines the energy emitted from the surface. Higher emissivity means greater ability of emittance. And the difference of surface emittance may have great effect on surface energy balance, which finally shapes the LST difference of the surfaces.

- The difference of sub-surface soil properties has significant contribution to the LST difference on both sides. Surface temperature changes strongly relates the dynamics of such sub-surface soil properties as soil heat capacity, thermal conductivity, hydraulic conductivity, soil moisture movement and so on. Though the lithological structure on both sides can be viewed as identical, the different process of surface energy balance due to different albedo may also lead to the difference in sub-surface soil properties, which in turn shapes an important impact on the LST difference on both sides.

Through the micro-meteorological simulation, these hypotheses can be tested on the surfaces of sand and biogenic crust. The mechanism leading to the simulation results on the two typical surface patterns can then be extended to understand the surface energy balance for an explanation of the LST difference on both sides of the region.

1.3 The methodology of the study

This study will adopt the combination of remote sensing analysis and micrometeorological modeling as its general methodology. The components involved in the study can be divided into two parts. (1) Remote sensing analysis of LST change on both sides and its validation through ground truth measurements. (2) Micrometeorological simulation for revealing the key factors contributing to the LST difference on both sides of the region.

Based on the availability, NOAA-AVHRR data and Landsat TM images will be used as the remote sensing data of this study. Therefore, how to estimate LST from the remote sensing data becomes one of the important aspects of the study.

The calculation of LST from NOAA-AVHRR data is generally through split-window algorithm. Price (1984) is one of those who firstly applied the technique to LST retrieval. Since Price (1984) a number of split-window algorithms have been proposed in order to improve the accuracy of LST estimation. Different algorithms were developed with different assumptions to the atmospheric effects and for different study purposes. I intend to establelish a new split window algorithm for the analysis of LST change on both sides of the region from NOAA-AVHRR data.

In order to compare LST change on both sides, a rectangle subset will be taken from the processed images on each side of the region. The rectangle subsets are set to be close enough to the border on both sides to meet the assumption of similar meteorological events on both sides. The size of the subsets is with 11×14 pixels, covering an area of about 12 km ×15km. Taking the smaller subsets than the whole region is with consideration of having relatively homogeneous hence minimizing the possible heterogeneous surface features for each side. The position of the subsets on each side is shown in Figure 1.4, which is a thermal image of the sand dune region, retrieving from NOAA-AVHRR-14 data acquired at 12:08 GMT or 15:08 local time on July 18, 1998 with a zenith angle of viewing 83.7°. The subsets are placed on the area that can maximally represent the general surface features of each side. The average LST of the subsets will be used as the representation of the LST change on both sides in the following analysis of the study.

Landsat TM thermal band (TM6) has a high spatial resolution (pixel size 120 m×120 m), which is suitable for detailed examination of small-scale LST patterns. No algorithm has been published for retrieval of LST from Landsat TM6 data even though the brightness temperature can be calculated using the method developed by the National Aeronautics and Space Administration (NASA) (Markham and Barker, 1986). In order to use this remote sensing data for the analysis, an algorithm for LST retrieval has to be developed in the study.

Micro-scale examination of the typical surfaces can be done through micro-meteorological modeling on the basis of surface energy balance theory. In the soil-atmosphere system, surface temperature is the direct indication of surface energy balance dynamics. In order to reveal the key factors governing the LST difference on both sides, it is necessary to model the performance of surface energy balance on the two typical surface patterns with respect to LST change. The model and its numerical solution will be described with details in chapters 6.

The micro-meteorological data about the region are needed for simulation and comparison. The required data includes global radiation, air temperature, wind velocity, emissivity, soil component and density, soil water content, air relative humidity. Some of these data are already available. For those that are not yet available, I intend to carry on my own measurements.

2 Remote sensing of land surface temperature

LST is an important factor controlling most physical, chemical and biological processes of the Earth. The knowledge of LST is strongly needed for many environmental studies and management activities of the Earth's resources (Asrar, 1989; Calvet, 1996). It is for the purpose of monitoring macro-scale spatial changes of surface temperature that scanners designed for sensing in the thermal bands are placed on board of platforms for remote sensing of the Earth's resources from space (Curran, 1985; Sabins, 1987; Drury, 1990).

The extensive application and significant importance of temperature in environmental studies and management is the main force promoting the study of LST in remote sensing. Under the availability of thermal sensing data such as channels 4 and 5 of AVHRR data as well as Landsat TM6, the study of the LST has become one of the hottest topics in remote sensing during the last decades (Vogt, 1996). With the development of remote sensing science, various algorithms have been proposed for the remote sensing of the LST. Different algorithms and methods are based on different considerations and are suitable for different conditions. Comparison of the existing algorithms is necessary so that a methodology of LST retrieval can be developed for the study.

Since the AVHRR data will be the main source of remote sensing data used for the study, the focus of this chapter is given to the progress in the researches on the remote sensing of LST using AVHRR data during the last two decades.

2.1 Theoretical basis for the remote sensing of LST

Land surface temperature is generally defined as the skin temperature of the ground. For the bare soil surface, LST is the soil surface temperature. However, the ground of the Earth is far from a skin or homogenous surface with two dimensions. Usually, it is composed of various objects on the surface and some of them such as vegetation may be best described in three dimensions. This situation makes the understanding of LST obscure, such as the case in a vegetated ground. The remote sensing of LST is based on the thermal spectral (long wave) radiation from the ground. Thus, LST in a dense vegetated ground can be viewed as the canopy surface temperature of vegetation and in a sparse vegetated ground, it is the average temperature of vegetation canopy, vegetation body and the soil surface under the vegetation (Calvet, 1996). Another factor leading to the difficult understanding of LST is that the surface is not homogenous under the spatial resolution (pixel scale) of remote sensing data. Usually, LST changes obviously in a small distance, such as 1m (Ottlé, 1992). However, the spatial resolution of most remote sensing data for LST is low compared with the difference of LST on the ground. For example, the pixel size under nadir is 1.1 km×1.1 km for NOAA-AVHRR and 120 m×120 m for Landsat TM channel 6. Thus, LST in remote sensing means the average surface temperature of the ground under the pixel scale mixed with different fractions of surface types (Kerr et al., 1992).

The theoretical basis for the remote sensing of LST is that total radiative energy emitted by ground surface increases rapidly with the increase of temperature. The spectral distribution of the energy emitted by ground objects also varies with temperature. According to Wien's Displacement Law on the relationship between spectral radiance and wavelength, for the Earth with an ambient temperature of 300K, the peak of its spectral radiance occurs at about 9.6 μm (Lillesand and Kiefer, 1987). Therefore, theoretically the thermal energy in relation to the physical temperature of the ground surface can be remotely observed by using sensors operating at wavelengths around 10 μm, which have been defined as thermal channels in remote sensing systems. The obtained temperature of the ground surface of the satellite level is called the brightness temperature (Reutter et al., 1994).

On the other hand, the spectral characteristics of the atmosphere indicate that there is an atmospheric window in the spectral region 8-14 μm, where atmospheric absorption is minimal and through which the thermal energy of ground surface can transmit without great losses. Thus, the ground surface temperature can be remotely sensed using the channels within the atmospheric window. Data from NOAA-AVHRR channels 4 and 5 operating respectively at 10.5-11.3 μm and 11.5-12.5 μm as well as Landsat TM channel 6 operating at 10.4-12.5 μm can be used to estimate surface temperature.

2.1.1 Plank's radiation equation and brightness temperature

In accordance with the blackbody radiation principle, the spectral radiance emitted from the ground surface as a blackbody can be described by Planck's radiance function (Humes et al., 1994; Saraf et al., 1995; Vogt, 1996):

$$B_\lambda(T)=\frac{c_1}{\lambda^5(e^{c2/\lambda T}-1)} \tag{2.1}$$

where $B_\lambda(T)$ is the spectral radiance, generally measured in W m^{-2} sr^{-1} μm^{-1}, λ is wavelength in m, c_1 and c_2 are the first and the second spectral constants respectively, $c_1=5.955215\times10^{-17}$ Wm^2 and $c_2=1.43876869\times10^{-2}$ m K.

If there is no attenuation in the process of transferring the emitted spectral radiance from the surface through the atmosphere to the remote sensor, the temperature of the ground object can be theoretically determined, if the emitted spectral radiance $B_\lambda(T)$ is measured, by inverting Planck's radiance equation as follows:

$$T=\frac{c_2}{\lambda\ln\left(\dfrac{c_1}{\lambda^5 B_\lambda(T)}+1\right)} \tag{2.2}$$

However, the spectral radiance of the ground surface is usually measured by a sensor installed in an airplane or satellite far from the ground. Therefore, the transmission of the emitted spectral radiance through the atmosphere to the sensor is affected by a number of factors, which make the retrieval of ground surface temperature of the remote sensing data more complicated. In the thermal wavelength, usually the atmosphere has three very important effects on the thermal radiance transmission: absorption, upward atmospheric radiance and bi-directional reflection of the downward atmospheric radiance (Vidal, 1991; Wooster et al., 1995). Therefore, the spectral radiance reaching the sensors is not only that emitted by ground objects and attenuated by atmospheric absorption, but also includes the radiance emitted by the atmosphere and the reflected component of the downward atmospheric radiance. Moreover, blackbody is a theoretical concept stating that the object has a full emission of its thermal radiance according to its temperature. Most objects on the Earth's surface do not behave as blackbody. The ratio between the radiation emitted by an object and that a blackbody with the same temperature is defined as the object's emissivity, which must be considered in remote sensing of the LST. Besides, different viewing angles of the sensor and the characteristics of the ground objects also have significant effects on the observed radiance of space (Ignatov and Dergileva, 1995; Wan and Dozier, 1996).

2.1.2 The satellite-observed radiation

Retrieval of LST from AVHRR having two thermal channels is possible through the application of so called split window algorithms derived from the radiance transfer equation of thermal wavebands. Usually, the formation of the equation is based on the following assumptions: (1) the Earth's surface is a Lambertian body; (2) a cloud-free atmospheric condition exists and (3) local atmospheric thermodynamic equilibrium conditions prevail. Therefore, the remotely-sensed radiance $B_i(T_i)$ for a given brightness temperature (T_i) in channel i is viewed at zenith angle θ and has undergone attenuation by atmospheric absorption, emission and scattering on its path from the surface of the Earth to the altitude of the satellite sensor which usually is in space at several hundred kilometers from the ground (França and Cracknell, 1994).

Different considerations to the radiation equation and different simplification to the atmospheric effects have resulted in different algorithms for the retrieval of the LST. According to Price (1984), the satellite-observed radiation $B_i(T_i)$ on top of the atmosphere can be expressed as

$$B_i(T_i)=B_i(T_s)-\int_0^{\tau_i}\exp(-\tau_i')\{B_i(T_s)-B_i[T_a(\tau_i-\tau_i')]\}d\tau_i' \tag{2.3}$$

where $B_i(T_s)$ is the ground surface radiation as a blackbody, T_s is LST, $B_i(T_s\ (\tau_i\text{-}\tau_i'))$ is the atmospheric radiation at the atmospheric temperature $T_a(\tau_i\text{-}\tau_i')$ which is a function of optical thickness $(\tau_i\text{-}\tau_i')$, in which τ_i and τ_i' are the optical thickness from ground and atmospheric height respectively to the sensor. Therefore, the second term of equation (2.3) represents the total atmospheric effects. Obviously, this equation treats these effects (absorption, upward and downward emitted radiance) as a negative term attenuating the emitted radiance by the ground.

This equation can not be directly solved because it contains several unknowns. Thus, simplification is necessary for its solution. Price (1984) neglected aerosol scattering in the 10-13 μm waveband and assumed the ground surface to be a blackbody. Thus, he only emphasized on the absorption and re-emission of radiation by atmospheric water vapor, which has the most important impact on the spectral transmission of thermal energy through the lower atmosphere. Then he used Taylor's expansion to linearize Planck's equation and only kept its first-order terms. Defining the total effect of atmospheric conditions as A_iU_t and assuming the optical thickness τ_i<<1 at the thermal band Price (1984) got:

$$T_i=(1-\tau_i)T_s+A_iU_t=T_s+A_iU_t \tag{2.4}$$

A split window algorithm for LST can be solved after applying this equation to channels 4 and 5 of AVHRR data for eliminating the atmospheric determinant U_t.

In order to formulate a split window algorithm for LST, Coll et al. (1994a) expressed the thermal radiance $B_i(T_i)$ obtained by the satellite scanner in channel i as follows:

$$B_i(T_i)=\varepsilon_i\tau_i(\theta)B_i(T_s)+I_i^{\uparrow}(\theta)+(1-\varepsilon_i)\tau_i(\theta)\gamma_iI_i^{\downarrow}(\theta=0) \tag{2.5}$$

where $\tau_i(\theta)$ is total atmospheric transmittance, θ is zenith angle of viewing (ZAV), $I_i^{\uparrow}(\theta)$ is the upward atmospheric radiance, $L_i^{\downarrow}(\theta=0)$ is the downward atmospheric radiance, γ_i is a parameter depending on both channel and type of atmosphere.

For derivation, several assumptions have to be made. A blackbody surface is generally assumed. Thus, only the atmospheric emission and absorption are accounted for. Besides, it is necessary to define a mean radiation temperature T_a to which the atmospheric radiance can be referred. Finally, Planck's function can be simplified by linearization in terms of temperature. By using the Taylor's expansion to the Planck function and only taking the first order of the expansion, equation (2.5) can be written with a good approximation as

$$T_s-T_i=\{[1-\tau_i(\theta)]/\tau_i(\theta)\}(T_i-T_a)+b_i(1-\varepsilon_i)/\varepsilon_i \tag{2.6}$$

where b_i has dimensions of temperature and can be approximated as follows:

$$b_i=\frac{T_i}{n_i}+\gamma_i(\frac{n_i-1}{n_i}T_i-T_a)[1-\tau_i(0)] \tag{2.7}$$

in which $\tau_i(0)$ is the atmospheric transmittance at nadir and n_i is a radiometric parameter which depends on channel and temperature interval considered and can be approximated as $B_i(T)\cong c_iT^{ni}$, in which c_i and n_i are constants for channel i. Given temperature range 280–320K, n_i=4.58972 for 8-14 μm (França and Cracknell, 1994) and c_i=1 for 10-12 μm (Price, 1983). Specifically, Schmugge et al. (1991) gave n_4=4.822

and n_5=4.472 for channel 4 and 5 of AVHRR while França and Cracknell, (1994) defined them as n_4=4.51921 and n_5=4.12636. Applying equation (2.7) to channels 4 and 5 of AVHRR for eliminating the term T_a, a split window algorithm can be derived.

Two atmospheric correction models have been developed by França and Cracknell (1994) for the retrieval of LST of remote sensing data in the thermal wavebands: Airborne model for data obtained by the sensor with low-flying platforms and split window algorithm for AVHRR data. For the later, the following radiance transfer equation has been used for the derivation on the basis of assumption that spectral emissivity and transmittance are nearly independent of temperature (França and Cracknell, 1994):

$$B_i(T_i) = \varepsilon_i \tau_i(\theta,0,Z) B_i(T_s) + \int_0^Z B_i(T_{az}) \frac{\partial \tau_i(\theta,z,Z)}{\partial z} dz + \\ 2\tau_i(1-\varepsilon_i)\int_0^{\pi/2}\int_\infty^0 B_i(T_{az}) \frac{\partial \tau'_i(\theta',z,0)}{\partial z} \cos\theta' \sin\theta' dz d\theta' \tag{2.8}$$

where Z is altitude of the sensor and z is the atmospheric altitude between the ground and the sensor, $\tau_i(\theta,0,Z)$ is atmospheric transmittance between the ground and the sensor, θ' is zenith angle of the downward atmospheric radiance, $B_i(T_{az})$ is atmospheric radiance for atmospheric temperature T_{az} at altitude z, $\tau_i(\theta,z,Z)$ is the atmospheric transmittance between z and Z, and $\tau'_i(\theta',z,0)$ is the atmospheric transmittance between z and the ground.

The meaning of each term at the right hand side of equation (2.8) is almost the same as that in the equation (2.8) though their expression of details and accuracy are quite different. The first term of right hand side in the equation (2.8) represents the ground radiance. The second term represents the upward atmospheric radiance and the third term represents the downward atmospheric radiance that is reflected by the surface and then attenuated in its upward path to the sensor.

After some rational assumptions for the whole atmosphere have been made for simplification and Taylor's expansion have been applied to Planck's function, the above radiance transfer equation becomes

$$T_s - T_i = \frac{1-\varepsilon_i}{\varepsilon_i} L_i + \frac{T_i - T_a}{\varepsilon_i \tau_i(\theta)\cos(\theta)} W_i - 2\frac{1-\varepsilon_i}{\varepsilon_i}(T_a - T_i + L_i) W_i \tag{2.9}$$

where $W_i = a_{1i}(\theta)w + a_{2i}(\theta)w^2$, in which w is water vapor content in the atmosphere and a_{1i} and a_{2i} are parameters about atmospheric state. T_a represents effective mean atmospheric temperature and L_i is given by $L_i = T_i/n_i$, in which n_i is a constant as in equation (2.7). Generally, a_{1i} and a_{2i}, are determined by simulation with LOWTRAN 7, a program for atmospheric simulation. A split window algorithm can be derived after equation (2.9) is applied to channels 4 and 5 of AVHRR data for eliminating the term T_a.

2.2 Split-window algorithms for LST retrieval

Because of its easy accessibility, NOAA−AVHRR data have been extensively applied to the examination of LST for different purposes. A number of split window algorithms for retrieval of LST from AVHRR data have been proposed in last decades. Though these algorithms differ greatly in calculation, two categories can be identified according to the form of their expression.

2.2.1 General form of split window algorithm

If T_4 and T_5 are brightness temperatures of AVHRR channels 4 and 5 respectively, the general form of the split-window equation can be written as

$$T_s = T_4 + A(T_4 - T_5) + B \tag{2.10}$$

where coefficients A and B are determined by impact of atmospheric conditions and other related factors on the thermal spectral radiance and its transmission through the atmosphere.

The split-window algorithm developed by Price (1984) for NOAA-7 has been extensively quoted in LST study. After making several simplifications to the atmospheric effects on the radiation transmission from the ground to the sensor, Price (1984) got his split window algorithm as (2.10), but he gave coefficient B=0 and coefficient A as

$$A = 1/(A_r - 1) \tag{2.11}$$

where $A_r = A_5/A_4$ is the function of atmospheric conditions, especially the contributions of atmospheric radiation and absorption in the thermal region. Successful use of the algorithm relies on the accuracy of determining the ratio A_r as a result of atmospheric effects. However, due to lack of detailed data on atmospheric profile *in situ* satellite pass, it is difficult to directly compute A_r.

It is assumed that LST is uniform in a small area with uniform surface features. Thus, the observed variation of brightness temperature is due to the atmospheric non-uniformity. Based on this assumption, Price (1984) estimated $A_r=\Delta T_5/\Delta T_4$, in which ΔT_4 and ΔT_5 are the average variation of brightness temperature in a small area with a relatively uniform surface feature on the images of channels 4 and 5 respectively. Price (1984) found that the ratio A_r=1.30 can be rational for most ground surfaces. Thus, equation (2.11) becomes $T_s=T_4+3.33(T_4-T_5)$. Obviously, this formula, if applying to other regions, may subject to possible errors due to the difference from the small area used to compute A_r and many unknown atmospheric effects.

In order to improve the accuracy of split-window algorithm, several modifications have been published since the mid-1980s. Based on their radiance transfer equation (2.5), which considers viewing angle and surface emissivity, Coll et al. (1994a) developed their split window algorithm. Though the algorithm keeps the general form of (2.10), its coefficients were given quite differently due to different considerations to atmospheric effects and ground emissivity:

$$A=[1-\tau_4(\theta)]/[\tau_4(\theta)-\tau_5(\theta)] \tag{2.12a}$$

$$B=\frac{1-\varepsilon_4}{\varepsilon_4}b_4+A\tau_5(\theta)[\frac{1-\varepsilon_4}{\varepsilon_4}b_4-\frac{1-\varepsilon_5}{\varepsilon_5}b_5] \tag{2.12b}$$

In which b_4 and b_5 can be given by equation (2.7) for channels 4 and 5. Obviously, coefficient A is only the function of atmospheric conditions. This is the same as in Price (1984). However, the coefficient B given by Coll et al. (1994a) is related to both atmospheric effects and ground emissivity, while it is equal to 0 in Price (1984).

Coll et al. (1994a) calculated the two coefficients A and B for a set of atmospheric profiles in their study. They found that water vapor content in the atmosphere, when it is above 1 g/cm^2, plays the most important effect in determining A, which is estimated to be 2.5-2.8 for the water vapor content 1–3.5 g/cm^2. After analyzing the impact of atmospheric transmittance on A, Coll et al. (1994b) proposed a modified equation to their algorithm. The modified equation still has the general form of equation (2.10), but coefficient A is determined as $A=1.29+0.28(T_4-T_5)$ and coefficient B is the same as the equation (2.12b). Obviously, coefficient A in the modified equation is not a constant, but depends linearly on the brightness temperature difference between channels 4 and 5 of AVHRR data. This is quite different from (18) and (2.12a).

Based on the structural components of ground surfaces, a methodology has been developed by Sobrino et al. (1991) for the atmospheric and emissivity corrections to the brightness temperature of AVHRR data. Though the algorithm of Sobrino et al. (1991) also has the general form as equation (2.10), its two coefficients A and B were given in a much more complicated form than the algorithms of Price (1984) and Coll et al. (1994a). The effects of surface emissivity, atmospheric absorption and water vapor content are directly expressed in the determination of coefficients A and B by Sobrino et al. (1991):

$$A = \frac{C_5 D_4 + C_4 D_5 w}{C_4 D_5 - C_5 D_4} \tag{2.13a}$$

$$B = \frac{(1-\varepsilon_4) C_4 D_5}{\varepsilon_4 (C_4 D_5 - C_5 D_4)}(1-2k_4 w)L_4 - \frac{(1-\varepsilon_5) C_5 D_4}{\varepsilon_5 (C_4 D_5 - C_5 D_4)}(1-2k_5 w)L_5 \tag{2.13b}$$

where w is water vapor content in the atmosphere (g/cm^2), k_i is atmospheric absorption coefficient in terms of water vapor content, C_i=$\varepsilon_i\tau_i\cos(\theta)$, D_i=$k_i[1+2\tau_i(1-\varepsilon_i)\cos(\theta)]$, L_i=T_i/n_i in which n_i is constant as in equation (2.7). Compared to Price (1984) and Coll et al. (1994a), this algorithm directly relates both coefficients A and B to the effects of atmospheric conditions and ground emissivity. By Price (1984) only atmospheric effects are considered for coefficient A and in Coll et al. (1994a) only coefficient B is related to both effects.

Based on their radiation transfer equation (2.8) and simplifications to the atmospheric effects, França and Cracknell (1994) also developed a split window algorithm for LST retrieval. The algorithm keeps the general form of the equation (2.10) with the two coefficients given as:

$$A=(D_5 C_4+D_4 C_5)/(D_5 C_4-D_4 C_5) \tag{2.14a}$$

$$B = \frac{(1-\varepsilon_4)(1-2W_4)L_4 D_5 C_4}{\varepsilon_4 (D_5 C_4 - D_4 C_5)} - \frac{(1-\varepsilon_5)(1-2W_5)L_5 D_4 C_5}{\varepsilon_5 (D_5 C_4 - D_4 C_5)} \tag{2.14b}$$

Where C_i= $\varepsilon_i\tau_i(\theta)\cos(\theta)$ and D_i= $W_i[1+2(1-\varepsilon_i)\tau_i(\theta)\cos(\theta)]$, in which W_i is defined in equation (2.9). This algorithm is very similar with Sobrino et al. (1991). Actually, C_i in (2.14) is the same as in (2.13). The only difference between them is that the viewing angle is considered for atmospheric transmittance in (2.14) while it is not in (2.13). D_i in (2.14) is also almost the same as in (2.13). The only difference is in the computation of atmospheric parameter W_i and k_i. Thus, it can be concluded that equation (2.14) are almost the same as (2.13).

Using the parameters A_0, P and M given by Becker and Li (1990), Sobrino and Caselles (1991) proposed a simplified algorithm with the two coefficients A and B as:

$$A=(M-P)/2 \tag{2.15a}$$

$$B=A_0+T_4(P-1) \tag{2.15b}$$

where the parameters A_0, P and M have been calculated by Becker and Li (1990) as $A_0=1.274$, and

$$P=1+0.15616(1-\varepsilon)/\varepsilon-0.482\,\Delta\varepsilon/\varepsilon^2 \tag{2.16a}$$

$$M=6.26+3.98(1-\varepsilon)/\varepsilon+38.33\,\Delta\varepsilon/\varepsilon^2 \tag{2.16b}$$

In which $\varepsilon=(\varepsilon_4+\varepsilon_5)/2$ and $\Delta\varepsilon=\varepsilon_4-\varepsilon_5$. As we can see, the coefficients A and B in the algorithm are directly expressed as the function of surface emissivity. It seems that atmospheric effects are given as constants. This emphasis is quite different from Price (1984), Coll et al. (1994a), Sobrino et al. (1991) and França and Cracknell (1994).

Using equation (2.10), theoretically, it should be possible to retrieve LST if atmospheric transmittance τ_i, water vapor content w for a given atmospheric profile and ε_4 and ε_5 of the surface are known. However, the detailed data for these parameters are usually not available *in situ* satellite pass. Thus, the method for estimating A and B is usually based on the simulation through LOWTRAN 7 (Coll et al., 1994a, Sobrino et al., 1991, França and Cracknell, 1994)

2.2.2 Other forms of split window algorithm

Apart from the general form of equation (2.10), other forms of split window algorithm for retrieval of LST from AVHRR data have also been proposed by different scholars in the last decades (Vogt, 1996; Norman et al., 1995; Badenas and Caselles, 1992). The algorithm given by Becker and Li (1990) is worthy of some detailed description since it has been cited in many papers (Givri, 1995; França and Cracknell, 1994) on LST study. Becker and Li (1990) presented a local split-window algorithm for viewing angles of up to 46° from nadir as follows:

$$T_s=A_0+P(T_4+T_5)/2+M(T_4-T_5)/2 \tag{2.17}$$

Where A_0, P and M are the same as in (2.15). Later Becker and Li (1995) modified their algorithm into a general one, but still keeping the form of (2.17). The only difference in the modified algorithm is that the coefficients are determined in terms of water content calculated from radiance simulation using LOWTRAN 7 program (Caselles et al., 1997).

Based on Becker and Li (1990), a generalized split window algorithm has been developed by Wan and Dozier (1996). The algorithm has the same form as (2.17) but the coefficients P and M are given as:

$$P=A_1+A_2\,\Delta\varepsilon/\varepsilon^2+A_3\,(1-\varepsilon)/\varepsilon \tag{2.18a}$$

$$M=B_1+B_2\,\Delta\varepsilon/\varepsilon^2+B_3\,(1-\varepsilon)/\varepsilon \tag{2.18b}$$

where A_1~A_3 and B_1~B_3 are parameters estimated by the method of Becker and Li (1990). The difference is that Wan and Dozier (1996) defined A_1 in their model as a variable while Becker and Li (1990) gave it as a constant equal to 1. Similarly, the parameters are determined by atmospheric simulation with LOWTRAN 7. Wan and Dozier (1996) compared the error of the generalized algorithm by the viewing angle θ-independent and θ-dependent methods. The simulation shows that the θ-dependent method is better than the θ-independent one. It can give an accuracy of 1K in LST retrieval.

After analyzing the atmospheric and emissivity efforts on LST determination, Prata (1993) developed an approximate split window algorithm from thermal radiance transfer equation with a full consideration of possible factors impacting the thermal radiance transfer through the atmosphere. The algorithm has a linear form as follows:

$$T_s=a_0+a_1T_4+a_2T_5 \tag{2.19}$$

where a_0-a_2 are coefficients representing total effects of atmospheric transmittance and ground emissivity. After applying the first-order Taylor expansion to approximate Planck's function, Prata (1993) derived the coefficients of his algorithm from his thermal radiance transfer equation as follows:

$$a_1=\frac{1+\gamma}{\varepsilon_4(1+\gamma\tau_5\Delta\varepsilon/\varepsilon_4)} \tag{2.19a}$$

$$a_2=\frac{\gamma}{\varepsilon_5(1+(1+\gamma)\tau_4\Delta\varepsilon/\varepsilon_5)} \tag{2.19b}$$

$$a_0=(1-a_1-a_2)[T_4-B_4(T_4)D]+(1-c)\Delta I^{\downarrow}D/c \tag{2.19c}$$

where $\Delta I^{\downarrow}=I_4^{\downarrow}-I_5^{\downarrow}$ is the downward atmospheric radiance difference between channel 4 and 5, $D=\partial B_4(T)/\partial T$ is derivative of Planck's function for temperature and the parameters are defined as $c=\varepsilon_4+\gamma\tau_5\Delta\varepsilon$ and $\gamma=(1-\tau_4)/(\tau_4-\tau_5)$, in which $\Delta\varepsilon=\varepsilon_4-\varepsilon_5$ is emissivity difference and τ_4 and τ_5 are atmospheric transmittance. As we can see, parameter a_1 and a_2 of the algorithm was determined as the function of ground emissivity and atmospheric transmittance. The calculation of parameter a_0 was very complicated, determined by parameter a_1 and a_2, down-welling atmospheric radiance, Planck's radiance of channel 4 and its derivative, and so on. Due to difficulty in estimating atmospheric radiance and calculating the derivative term, the application of this algorithm is also inconvenient. For simplification, Prata (1993) evaluated the variation of the coefficient a_0 according to the US standard atmosphere. He found that the coefficient changes linearly with temperature for specific emissivity and transmittance and that emissivity effects cause greater variation of the coefficient than temperature effects. For $\Delta\varepsilon$=-0.01 and ε_4=0.96, the coefficient a_0 decreases from 4.2°C to 2.8°C in the temperature range 0-50°C. Prata (1994) validated this algorithm through applying to the Melbourne region of Australia for comparison of LST retrieved from AVHRR and ground truth measurement in his experimental sites. The result indicated that the root-mean square difference between satellite-retrieved LST and the ground-measured temperature is about ±1.5°C.

Ottlé and Vidal-Madjar (1992) also presented a split window algorithm for LST retrieval from AVHRR, which has the same linear form as equation (2.19). It is well known that LST is extremely variable in space. Even within a distance of 1m, an obvious LST difference can be obtained. Moreover, it is not a simple matter to average ground truth data (if available) at the satellite pixel scale. In order to estimate the coefficients of equation (2.19), 1207 atmospheric situations from the Automatized Atmospheric Absorption Atlas have been considered by Ottlé and Vidal-Madjar (1992) for the three general air masses: tropical, mid-latitude and polar systems. Regressions were computed for each possible viewing angle. Table 2.1 presents the results for the mid-latitude atmosphere, by assuming that the surface behaves like a perfect blackbody in the two spectral bands. The result indicates that viewing angle has little influence on the coefficient value of the algorithm and the root mean square (RMS) error is very small in all the cases, increasing very slowly with angle.

It is well known that the radiance of thermal wavebands is strongly affected by surface emissivity. Ottlé and Vidal-Madjar (1992) found that an error of 2% on the mean emissivity could yield an error of 1K in LST estimation. This is much better than what Becker (1987) estimated δT=1K for $\delta\varepsilon$=1%. Generally, the ground emissivity can be assumed to be 0.96-0.98 for most surfaces. Under this assumption, Ottlé and Vidal-Madjar (1992) evaluated the effect of emissivity difference between channels 4 and 5 on LST estimation. The result shows that the effect of emissivity on LST error increases with scan angle in all cases. For an error of emissivity difference $\delta(\Delta\varepsilon)$=2%, δT is about 1.5K for nadir view and 2.5K for 53°. When using $\delta(\Delta\varepsilon)$=1%, they got δT to be about 1K for the same conditions as in Becker (1987) who predicted δT=2.5K for $\delta(\Delta\varepsilon)$=1%.

Table 2.1 Coefficients of the AVHRR split window algorithm (after Ottlé and Vidal-Madjar 1992)

Viewing angle θ (°)		Coefficients			RMS error (K)
		a_0	a_1	a_2	
0.0		0.858	3.218	-2.218	0.123
9.0		0.854	3.225	-2.225	0.123
23.0		0.852	3.258	-2.258	0.135
32.0		0.880	3.289	-2.290	0.145
44.0		0.928	3.372	-2.372	0.174
53.0		0.929	3.468	-2.469	0.211
Emissivity	θ	a_0	a_1	a_2	RMS error
0.98	0	-0.403	3.219	-2.211	0.111
0.98	53.0	-0.418	3.506	-2.499	0.201
0.96	0	-1.687	3.213	-2.197	0.102
0.96	53.0	-1.761	3.487	-2.471	0.184
ε_4=0.98	0	-0.502	3.023	-2.013	0.116
ε_5=0.985	53	-0.515	3.349	-2.339	0.201
ε_4=0.96	0	-2.186	2.444	-1.420	0.173
ε_5=0.98	53	-2.239	2.830	-1.804	0.250
ε_4=0.98	0	-1.301	2.510	-1.492	0.161

Kerr et al. (1992) also proposed an algorithm for retrieval of LST in semi-arid and arid environments. The algorithm is similar to equation (2.19). They considered the surface in arid and semi-arid environment as composed of vegetation coverage and the bare soil though vegetation cover is sparse. According to equation (2.19), the surface temperatures of vegetation canopy T_v and bare soil T_{bs} can be estimated respectively. Then, they considered LST as a function of T_v and T_{bs} according to the following formula:

$$T_s=R_vT_v+(1-R_v)T_{bs} \tag{2.20}$$

Where R_v is vegetation cover rate in semi-arid and arid environment, which can be derived from the normalized difference of vegetation index (*NDVI*) as follows (Kerr et al., 1992):

$$R_v=(NDVI-NDVI_s)/(NDVI_v-NDVI_s) \tag{2.21}$$

Where $NDVI_s$ is the minimum *NDVI* for bare soil and $NDVI_v$ the highest *NDVI* of a fully vegetated pixel. The unique characteristic of this algorithm is that it divides the surface into two typical patterns and considers their various contributions respectively to LST in pixel scale while others view the surface as one pattern. Another unique feature is that this algorithm directly relates LST to *NDVI*, an important index measuring vegetation on the ground surface.

The algorithm presented by Kerr et al. (1992) has been tested and validated over two sites. One is located in the southeast of France (Caumont near Avignon 43°54'N-4°51'E) and the other is in Niger in Dangurey Gourou (13°59'N-2°E) near Niamey. They got the empirical solution of the coefficients in equation (2.19) as a_0=3.1, a_1=3.1 and a_2=-2.1 for bare soil and a_0=-2.4, a_1=3.6 and a_2=-2.6 for vegetation. For the computing R_v they used $NDVI_s$=0.11 and $NDVI_v$=0.72. The result shows that the algorithm fits rather well with the measured data. The difference between estimated and measured LST has a mean value of –0.25K and a standard deviation of 0.9K (Kerr et al., 1992). This accuracy is much better than the general accepted error 1.5 K. Because the arid land environment is the most difficult case for split window technique, Kerr et al. (1992) believed that this algorithm can give a better result to other land surfaces.

2.3 Effect of ground emissivity on retrieval of LST

When applying the split window algorithm to retrieve LST from AVHRR data, one potential complication arises from the possible effects of emissivity variation of natural surfaces in the thermal window. The derivation of many algorithms for LST is based on the assumption that the ground surface acts as a blackbody with ε=1. Actually, the Earth's surface is not a blackbody. All materials emit only a fraction of the energy emitted by a blackbody at the same temperature. Therefore, the thermal radiance received by the sensors on board satellite is also determined by surface emissivity (Barducci and Pippi, 1996). The emitting ability of a ground object, compared to that of a blackbody, is referred to as the object's emissivity, which describes how efficiently the object emits its thermal energy compared to a blackbody. When using AVHRR data to calculate LST, it is important to know the emissivity of the surface and how it affects on LST.

2.3.1 Emissivity correction to split window algorithms

Becker (1987) evaluated the possible error δT of LST estimation generated by the surface emissivity as follows

$$\delta T \cong 50(1-\varepsilon)/\varepsilon - 300\Delta\varepsilon/\varepsilon \tag{2.22}$$

where $\Delta\varepsilon=\varepsilon_4-\varepsilon_5$ is the difference of ground emissivity in the two AVHRR channels. Assuming the average emissivity to be $\varepsilon=0.96$ and the difference of emissivity to be $\Delta\varepsilon=0.002$, the probable LST error δT would be 1.4583K. Therefore, δT may be quite significant when the ground surface is obvious heterogeneity in the study region. And this is generally the case for natural surfaces.

Li and Becker (1993) predicted δT due to emissivity error as $\delta T \cong -52\delta\varepsilon - 110\delta(\Delta\varepsilon)$, in which $\delta\varepsilon$ is the error of ε and $\delta(\Delta\varepsilon)$ is the error of $\Delta\varepsilon$. If $\delta\varepsilon=\delta(\Delta\varepsilon)=0.01=1\%$, δT would be larger than 1.6K. Therefore, it is necessary to have a good knowledge of surface emissivity in order to have a good LST estimation using the AVHRR data.

The surfaces in agricultural areas have an average emissivity of $\varepsilon=0.97$ in the thermal spectral interval 10.8-11.9 μm and a value of 0.96 maybe is a good assumption for most surfaces with vegetation cover (Price, 1984). According to the relation between emissivity and temperature, after accounting for the effect of surface emissivity variation, Price (1984) calibrated his algorithm as:

$$T_s=[T_4+3.33(T_4-T_5)][\frac{5.5+\varepsilon_4}{4.5}]+0.75T_5(\varepsilon_4-\varepsilon_5) \tag{2.23}$$

Thus, at the Earth's ambient temperature 300K a $\delta(\Delta\varepsilon)=0.01$ may causes a $\delta T=2$K.

Compared to Price (1984), Coll et al. (1994a) considered the effect of ground emissivity only in the coefficient B of their algorithm. Other algorithms such as Sobrino et al. (1991) and França and Cracknell (1994) directly relate the determination of both coefficients A and B to ground emissivity. In the algorithm presented by Becker and Li (1990) and Wan and Dozier (1996), ground emissivity is also a key factor for determining the required coefficients. Therefore, ground emissivity is very important for an accurate LST retrieval from AVHRR data.

2.3.2 Methods for determining the ground emissivity

Humes et al. (1994) carried on a number of experiments for estimating ground emissivity. They used the "cone method" to acquire surface radiant temperature T_r, surface kinetic temperature T_s and air temperature T_a. Then, Planck's equation was used to convert the three temperatures into their emittances $B(T_r)$, $B(T_s)$ and $B(T_a)$. The relationship among the three emittances can be expressed as

$$B(T_r)=\varepsilon B(T_s)+(1-\varepsilon)B(T_a) \tag{2.24}$$

When solved for the surface emissivity ε, equation (2.24) gives

$$\varepsilon=[B(T_r)-B(T_a)]/[B(T_s)-B(T_a)] \tag{2.25}$$

Humes et al. (1994) measured various surfaces for estimating ground emissivity. Their computation with a total of 183 experimental samples gave the results of mean emissivity for bare soil-rock mixture 0.959, shrub and clumsy vegetation 0.994 and rock-soil-vegetation mixture 0.981.

Surface emissivity in channels 4 and 5 must be calculated in determining the coefficients of split window algorithms such as Coll et al. (1994a), Sobrino et al. (1991), Sobrino and Caselles (1991) and Becker and Li (1990) and the calibration of the algorithms such as Price (1984). However, the appropriate filters of the radiometers for channels 4 and 5 are generally not available for the field truth measurement. By using the field radiometers operating in 8-14μm, a methodology has been created by Sobrino and Caselles (1991) to calculate the emissivity correlation of channels 4 and 5 as $\varepsilon_5=\varepsilon_{8\text{-}14}+0.001$, $\varepsilon_4=\varepsilon_{8\text{-}14}-0.003$ and $\Delta\varepsilon=\varepsilon_5-\varepsilon_4=0.004\pm0.002$. Experiments on several ground surfaces indicate that this method produces an estimation quite reasonable (Sobrino and Caselles, 1991).

Using the power law relation between the radiance and temperature $B_i(T_i)=c_iT^{ni}$, Schmugge et al. (1991) provided a simple way to estimate the approximate surface emissivity. Actually, $B_i(T_i)=\varepsilon_iB_i(T_s)$. Thus, taking the ratio of the two channels 4 and 5, one can obtain

$$\varepsilon_5=\varepsilon_4^{n5/n4}(T_5/T_4)^{n5} \tag{2.26}$$

For the same type of surface in an image, the emissivity ε_i may be relatively constant, but a large amount of variation in T_s may exist, which leads to the variation of T_i on the ground. By summing over m pixels for an average T_i, formula (2.26) can be used to estimate ε_5 when ε_4 is given.

It is well known that, in channel 3 of AVHRR data which is located around 3.7 μm, the radiance emitted by the land surface itself and the reflected radiance due to sun irradiation during the day are of the same order of magnitude. Using this very interesting property, Li and Becker (1993) proposed a method for determining surface emissivity from AVHRR data using only the sun as an active source. The general idea behind this method is that the emitted radiance during the night can be used to evaluate the emitted radiance during the day, thanks to the temperature independent spectral indices (*TISI*), which are defined by Li and Becker (1993) as

$$TISI_n = M_c \frac{B_3(T_{g3n})}{B_4(T_{g4n})^{m_c} B_5(T_{g5n})} \tag{2.27a}$$

$$TISI_d = M_c \frac{B_3(T_{g3d})}{B_4(T_{g4d})^{m_c} B_5(T_{g5d})} \tag{2.27b}$$

Where n and d respectively refer to night and day. Thus, $B_3(T_{g3n})$ represents the Planck's function in channel 3 with the ground temperature T_{g3n} during the night. M_c and m_c are constants that have been determined according to the procedure given in Becker and Li (1990). For AVHRR of NOAA-11, $M_c = 0.62105 \times 10^{-7}$ and $m_c = 1.924$. Following several complex logical derivations and approximations, the ground emissivity of channels 4 and 5 can be calculated as:

$$\varepsilon_4 = \{\frac{\varepsilon_3(\theta)}{TISI_{54n} TISI_n}\}^p \tag{2.28a}$$

$$\varepsilon_5 = \varepsilon_4^{n54}\, TISI_{54n} \tag{2.28b}$$

Where n_{54} and p are constants with $n_{54} = n_5/n_4 = 0.913$ and $p = 1/(n_{54} + m_c) = 0.352485$, $TISI_{54n}$ is a parameter determined as

$$TISI_{54n} = (T_{g5n}/T_{g4n})^{n5} \tag{2.29}$$

Where n_5 and n_4 are the same parameters as in equation (2.7). And

$$\varepsilon_3(\theta) = 1 - \{\pi \frac{(TISI_d - TISI_n) B_4(T_{g4d})^{m_c} B_5(T_{g5d})}{M_c R \tau_3(\theta') \cos(\theta')}\} \tag{2.30}$$

Where R is constant with $R = 0.415536 \times 10^{-6}\pi$ and $\tau_3(\theta')$ atmospheric transmittance in channel 3.

A method for estimating emissivity difference has been proposed by Coll et al.

(1994b). The method is based on the separation between the atmospheric and emissivity effects on the T_4-T_5. Taking into account several approximations for a thermal radiance transfer equation, Coll et al. (1994b) expressed T_s-T_i at the surface as

$$T_s - T_i = \frac{1-\varepsilon_i}{\varepsilon_i} b_i \tag{2.31}$$

Where b_i is a parameter given by equation (2.7). Applying equation (2.31) to channels 4 and 5 and eliminating T_s from the equation system, we gets

$$T_4 - T_5 = \frac{1-\varepsilon_5}{\varepsilon_5} b_5 - \frac{1-\varepsilon_4}{\varepsilon_4} b_4 \tag{2.32}$$

This relates the difference between the atmospherically corrected temperatures to the ground emissivity in both channels. When ε_i is close to 1 as is the case in 10.5~12.5μm for most natural surfaces, the above equation can be further simplified by using $\varepsilon_5 = \varepsilon_4 - \Delta\varepsilon$ to give

$$T_4 - T_5 = (1-\varepsilon_4)(b_5 - b_4) + \Delta\varepsilon b_5 \tag{2.33}$$

Solution of equation (2.33) for $\Delta\varepsilon$ gives

$$\Delta\varepsilon = [(T_4 - T_5) - (1-\varepsilon_4)(b_5 - b_4)]/b_5 \tag{2.34}$$

Provided that a priori estimate of ε_4 is available, this formula can be used to calculate $\Delta\varepsilon$. Sensitivity analysis indicates that the total error in $\Delta\varepsilon$ by this method is about 0.004 (Coll et al., 1994b). This accuracy is sufficient for most LST applications, since the corresponding error in LST estimation is less than 1K when the general form of the split window algorithm is employed.

Coll et al. (1994b) applied this method to the real data of an experiment in southern France. They used the radiosonde profile of the experiment performed in June 1992 to calculate the required atmospheric parameters. Consequently the coefficients b_i and the emissivity differences $\Delta\varepsilon$ can be estimated, given that $\varepsilon_4=0.98$ for forest land in the experimental area. The result indicates that the mean $\Delta\varepsilon$ is about 0.0081, which is higher than a general assumption $\Delta\varepsilon=0.002$. Thus, $\Delta\varepsilon$ has to be computed for an accurate LST retrieval using a split window algorithm.

2.4. Summary of the chapter

The extensive requirement of LST for the environmental studies and management activities of the Earth's resources has made the remote sensing of LST an important topic during the last two decades (Prata et al., 1995). Many studies have been devoted to establelishing the methodology for the retrieval of LST from channel 4 and 5 of AVHRR data. Various split window algorithms have been reviewed and compared in the chapter to understand their differences. Different algorithms differ in both their forms and the calculation of their coefficients. The most popular form of the algorithm is $T_s=T_4+A(T_4-T_5)+B$ though some other forms such as the linear one have also been used. The determination of coefficients A and B has great difference among various algorithms, in terms of the detailed derivations of the algorithms from the approximate assumptions to the different conditions and factors involved.

Price (1984) is one of those who first applied the split window technique to LST retrieval. The ground surface was assumed as a black body in his derivation even though he later altered his algorithm by adding a correction term of emissivity. Several modifications to the method of Price (1984) have been published since mid-1980s. Coll and Caselles (1994) added the satellite zenith observation θ and surface emissivity ε_i of channel i into the radiation transfer equation and modified the split window algorithm into a new form accordingly. Except transmittance and ground emissivity, this algorithm requires the known of other two atmospheric parameters for estimation of its coefficients: mean atmospheric temperature T_{ai} and the parameter γ_i, which they did not mention how to estimate. Based on the structural components of ground surface, a methodology has been developed by Sobrino et al. (1991) for atmospheric and emissivity corrections to the brightness temperature of AVHRR data. Ground emissivity, atmospheric transmittance and the other two parameters (water vapor content in atmospheric profile and the parameter stating about atmospheric absorption) are directly involved in their algorithm. Through a complicate radiation transfer equation, França and Cracknell (1994) establelished two atmospheric correction models for retrieval of LST from remote sensing data in thermal wavebands: airborne model for data obtained from low flying altitude and a split window algorithm for AVHRR data. Again, water vapor content and an atmospheric parameter were also involved in this algorithm. The general form of these split window algorithms is the same as mentioned in the previous paragraph, but the calculation of the coefficients A and B is totally different, due to different considerations to the atmospheric impacts on the radiation transfer through the air. Using a linear relationship between LST and brightness temperature, Ottlé and Vidal-Madjar (1992) presented a split window algorithm as $T_s=a_0+a_1T_4+a_2T_5$, in which the coefficients are determined through regression method. The problem of applying this algorithm in the real world is how to get a representation of LST for a set of corresponding pixels with a size of up to 1.1 km×1.1km. After analyzing the atmospheric and emissivity efforts on LST determination, Prata (1993) derived an approximate split window algorithm from thermal radiance transfer equation with a full consideration of possible factors. This algorithm has the same linear form as Ottlé and Vidal-Madjar (1992) but the calculation of its parameters are totally different. Parameter a_1 and a_2 were determined as the function of ground emissivity and atmospheric transmittance. The calculation of parameter a_0 was very complicated, determined by parameter a_1 and a_2, down-welling atmospheric radiance, Planck's radiance of channel 4 and its derivative, and so on. Due to difficulty in estimating atmospheric radiance and calculating derivative term, the application of this algorithm is also inconvenient. Another important version of split window algorithm is developed by Becker and Li (1990) with the form of $T_s=A_0+P(T_4+T_5)/2+M(T_4-T_5)/2$, where A_0, P and M are coefficients. For determination of the coefficients, they computed the atmospheric impacts as constant and gave more attention on the effect of ground emissivity. The assumption of atmospheric effects as constant is not true for most cases. This algorithm was further modified by Wan and Dozier (1996) as a viewing angle dependent method for achieving higher accuracy of LST retrieval.

Atmospheric transmittance and ground emissivity are believed to be the two most important factors impacting on the accuracy of LST retrieval from remote sensing data. Usually atmospheric transmittance can be estimated through simulation using such programs as LOWTRAN and MODTRAN. In the thermal portion of the spectrum, the ground emissivity is close to 1 even though different components of surface materials have different emissivity values. For an approximate estimation, it is usual to assume a ground emissivity of 0.96 for the application of the split window algorithm. Actually, this is not true because different surfaces have different emissivity, which may range from 0.91 to 0.99. Furthermore, the ground emissivity is not equal but a little bit difference in the two AVHRR channels. This difference has been proved to be very important for the accuracy of LST retrieval.

Various methods have been created to estimate ground emissivity and its difference in the two thermal channels (Hook et al., 1992, Beck and Li,1993). Theoretically, the ground emissivity in channels 4 and 5 can be derived from the remote sensing data of AVHRR channels 3, 4 and 5 based on some assumptions. However, the method also involves a number of atmospheric parameters, which are generally difficult to determine because the needed data are usually not available *in situ* the satellite pass. Therefore, the estimation of ground emissivity also requires the standard atmosphere profile provided by the LOWTRAN program to simulate the effects of the potential factors involved. Studies indicate that the ground emissivity of channels 4 and 5 may have a difference ranging 0.002-0.008 (Sobrino and Caselles, 1991; Coll et al., 1994b).

Most of the existing presentations of the split window algorithm do not give a detailed description of the derivation process of the algorithm. Some gave a simple derivation, but omitted many processes. The incomplete derivation makes the understanding of the existing split window algorithms, especially the computation of the coefficients, inconvenient and difficult. Moreover, the existing algorithms such as Sobrino et al. (1991), Coll and Caselles (1994) and França and Cracknell (1994) involve some parameters that are not easy to estimate when applying to the real world. Besides, sensitivity analysis of split window algorithm to its parameters is usually neglected in existing presentations of the methodology. This again causes difficulty for their application to the real world because in many applications the knowledge about probable LST estimation error due to possible error in parameter determination is generally desired so that assessment can be given to the accuracy of the estimation using the desired algorithm. In addition, the lack of sensitivity analysis in existing algorithms prevents such assessment when applying them to the real cases especially to a large-scale region. The current existing algorithms are not suitable for the study region. Therefore, in order to analyze the temporal and spatial LST change in the region, a proper methodology of LST retrieval has to be developed.

3 LST change from NOAA-AVHRR data in the region

The interesting phenomenon of LST anomaly on both sides of the border region is observed in the daytime remote sensing images. Application of remote sensing techniques to analyze the temporal and spatial change of the phenomenon is absolutely necessary before any attempt is taken to explain its occurrence. Due to availability, the NOAA-AVHRR and Landsat TM data were used for the analysis. AVHRR has two thermal channels operating in 10.5-11.3μm and 11.5-12.5μm, respectively. Though its spatial resolution is relatively low (1.1 km×1.1km under nadir), NOAA-AVHRR has the advances of high revisit time (about two images a day) and easy access (public domain data at many receiving stations). These two features made it become an important source of remote sensing data for monitoring the Earth's resources. Landsat TM has one thermal channel operating in 10.4-12.5μm with a spatial resolution of 120 m ×120 m. Thus, it is very suitable for analysis of spatial LST change. In order to use these two kinds of remote sensing data for LST analysis, first of all, methods have to be developed for retrieving LST from the remote sensing data. Then the methods are applied to the region for analysis of LST change.

3.1 Two-factors split window algorithm for LST retrieval

The retrieval of LST from NOAA-AVHRR data is mainly achieved through the application of the split window algorithm, though other algorithms such as the one establelished by Reutter et al. (1994) are also in use. The basic idea of the split window algorithm can be traced back to the expression of McMillin (1975) in an equation stating that for a given temperature, there is a linear relationship between the radiance and temperature in the two infrared windows. Since the two thermal channels of AVHRR are close enough to each other, but still with differences of transmittance and emissivity, an equation system about thermal radiance reaching the remote sensor in the two channels can be solved for temperature without the need of precise atmospheric temperature and pressure profiles.

3.1.1 Thermal radiance transfer from ground to remote sensor

The derivation of a split window algorithm for LST retrieval is based on thermal radiance of the ground and its transfer from the ground through the atmosphere to the remote sensor. Generally speaking, the ground is not a back body. Thus, the ground emissivity has to be considered in computing the thermal radiation emitted from the ground. the atmosphere has important effects on the received radiance at remote sensor level. Considering all these impacts, we have the general radiance transfer equation (Ottlé and Stoll, 1993; Cracknell and Xue, 1996) for remote sensing of LST as follows:

$$B_i(T_i)=\tau_i(\theta)[\varepsilon_i B_i(T_s)+(1-\varepsilon_i)I_i^{\downarrow}]+I_i^{\uparrow} \tag{3.1}$$

where T_s is LST and T_i is the brightness temperature in channel i, $\tau_i(\theta)$ is atmospheric transmittance at viewing direction θ (zenith angle from nadir) and ε_i is ground emissivity. $B_i(T_i)$ is radiance received by the sensor, $B_i(T_s)$ is ground radiance, $I_i^{\downarrow}$ and $I_i^{\uparrow}$ are down-welling and upwelling atmospheric radiance, respectively.

The upwelling atmospheric radiance $I_i^{\uparrow}$ is usually computed (França and Cracknell, 1994; Cracknell, 1997) as

$$I_i^{\uparrow}=\int_0^Z B_i(T_z)\frac{\partial\tau_i(\theta,z,Z)}{\partial z}dz \tag{3.2}$$

where T_z is atmospheric temperature at altitude z, Z is altitude of the sensor, $\tau_i(\theta,z,Z)$ represents the upwelling atmospheric transmittance from altitude z to the sensor height Z. Following McMillin (1975), Prata (1993) and Coll et al. (1994a), we employ the mean value theorem to express the upwelling atmospheric radiance as follows:

$$B_i(T_a)=\frac{1}{1-\tau_i(\theta)}\int_0^Z B_i(T_z)\frac{\partial\tau_i(\theta,z,Z)}{\partial z}dz \tag{3.3}$$

where T_a is the effective mean atmospheric temperature and $B_i(T_a)$ represents the effective mean atmospheric radiance with T_a in channel i. Thus, we get

$$I_i^{\uparrow}=(1-\tau_i(\theta))B_i(T_a) \tag{3.4}$$

The down-welling atmospheric radiance is generally viewed as from a hemispherical direction, hence can be computed as

$$I_i^{\downarrow}=2\int_0^{\pi/2}\int_{\infty}^0 B_i(T_z)\frac{\partial\tau'_i(\theta',z,0)}{\partial z}\cos\theta'\sin\theta'\,dz\,d\theta' \tag{3.5}$$

where θ'is the down-welling direction of atmospheric radiance and $\tau'_i(\theta', z, 0)$ represents the downwelling atmospheric transmittance from altitude z to the ground surface. Followed França and Cracknell (1994), it is rational to assume that $\partial\tau'_i(\theta', z, 0) = \partial\tau_i(\theta, z, Z)$ for the thin layers of the whole atmosphere. Based on this assumption, application of mean value theorem to the equation (3.5) gives

$$I_i^{\downarrow} = 2\int_0^{\pi/2} (1-\tau_i'(\theta))B_i(T_a)\cos\theta'\sin\theta' d\theta' \quad (3.6)$$

The integration of this equation can be solved as

$$2\int_0^{\pi/2} \cos\theta'\sin\theta' d\theta' = (\sin\theta')^2\Big|_0^{\pi/2} = 1 \quad (3.7)$$

Substituting the solution into the equation (3.6), we get

$$I_i^{\downarrow} = I_i^{\uparrow} = (1-\tau_i(\theta))B_i(T_a) \quad (3.8)$$

Thus, the radiance equation (3.1) can be rewritten as

$$B_i(T_i) = \varepsilon_i\tau_i(\theta)B_i(T_s) + (1-\tau_i(\theta))[1+(1-\varepsilon_i)\tau_i(\theta)]B_i(T_a) \quad (3.9)$$

This equation can be applied to the thermal channels 4 and 5 of AVHRR to give an equation system as follows

$$B_4(T_4) = \varepsilon_4\tau_4(\theta)B_4(T_s) + (1-\tau_4(\theta))[1+(1-\varepsilon_4)\tau_4(\theta)]B_4(T_a) \quad (3.10a)$$

$$B_5(T_5) = \varepsilon_5\tau_5(\theta)B_5(T_s) + (1-\tau_5(\theta))[1+(1-\varepsilon_5)\tau_5(\theta)]B_5(T_a) \quad (3.10b)$$

Based on this equation system, a split window algorithm can be derived under several assumptions and simplifications to the function of Planck's radiance in the equation.

3.1.2 Linearization of Planck's radiance function

In order to develop a split window algorithm from the equation (3.9), we need to apply the method of linearizing Planck's radiance function, which allows us to directly relate the radiance to the temperature in the corresponding channels. It is a critical step in deriving split window algorithm.

The linearization of Planck's radiance function can be done through the application of Taylor's expansion method. As indicated in Figure 3.1a, the relationship between Planck's radiance and temperature in the two AVHRR channels is almost linear for a given wavelength. In this case, keeping the first two terms of Taylor's expansion can achieve a high accuracy of the approximation. This is the general approximation for deriving split window algorithm (Price, 1983; França and Cracknell, 1994; Coll et al., 1994a). Thus, Planck's radiance function can be approximated by Taylor's expansion, for the first two terms, as

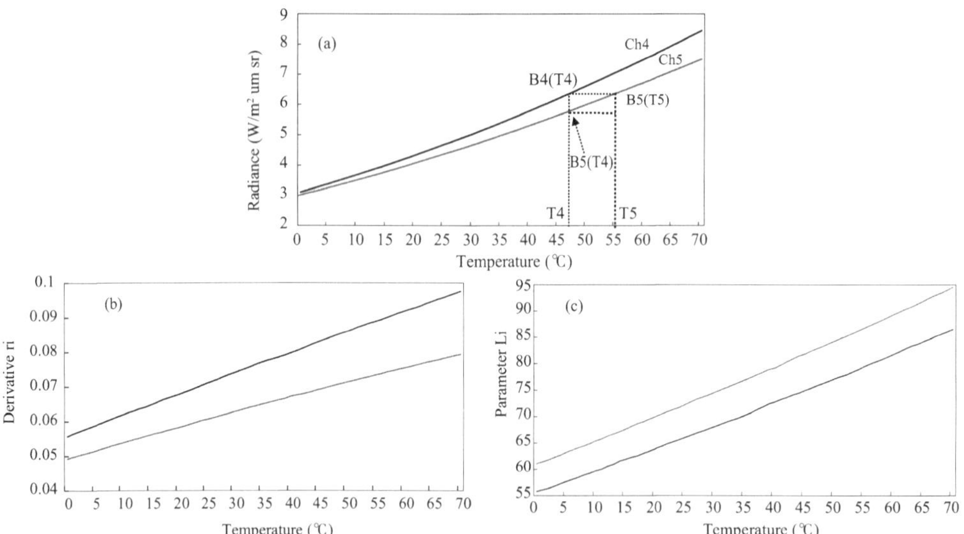

Figure 3.1 Linearization of Planck's function, plotting (a) Planck's radiance *vs.* temperature; (b) derivative of Planck's radiance *vs.* temperature; (c) parameter L_i *vs.* temperature.

$$B_i(T_j)=B_i(T)+(T_j-T)\partial B_i(T)/\partial T=(L_i+T_j-T)\partial B_i(T)/\partial T \tag{3.11}$$

where i refers to channel 4 or 5, T_j refers to the temperatures of channel 4 (j=4), channel 5 (j=5), land surface (j=s) and air (j=a). The parameter L_i is given as

$$L_i= B_i(T)/[\partial B_i(T)/\partial T] \tag{3.12}$$

In which L_i has the dimension of temperature in K. The physical meaning of Taylor's expansion in this case can be illustrated in Figure 3.1a: it expresses the radiance of $B_i(T_j)$ in terms of the radiance $B_i(T)$ with a fixed temperature T, which is usually defined as T_4 (França and Cracknell, 1994; Prata, 1993). For a moderate departure $|T-T_j| \leqslant 10$K, the accuracy is better than 1% for 10-13μm within the temperature range 270-320 K.

Thus, to express the Planck's radiance of channel 5 for brightness temperature T_5, we have

$$B_5(T_5)=B_5(T_4)+(T_5-T_4)\partial B_5(T_4)/\partial T=(L_5+T_5-T_4)\partial B_5(T_4)/\partial T \tag{3.13a}$$

Similarly, we get

$$B_4(T_4)=(L_4+T_4-T_4)\partial B_4(T_4)/\partial T=L_4\partial B_4(T_4)/\partial T \tag{3.13b}$$

$$B_4(T_s)=(L_4+T_s-T_4)\partial B_4(T_4)/\partial T \tag{3.13c}$$

$$B_4(T_a)=(L_4+T_a-T_4)\partial B_4(T_4)/\partial T \tag{3.13d}$$

$$B_5(T_s)=(L_5+T_s-T_4)\partial B_5(T_4)/\partial T \tag{3.13e}$$

$$B_5(T_a)=(L_5+T_a-T_4)\partial B_5(T_4)/\partial T \tag{3.13f}$$

For the parameter L_i, Slater (1980) and Price (1984) suggested the following approximation,

$$L_i=T_i/n_i \tag{3.14}$$

The values of n_i for the channel 4 and 5 are 4.592 and 4.12636, respectively, for the temperature range 280-320K (França and Cracknell, 1994). Actually, n_i is not a constant, but changes linearly as a function of temperature. Regression of n_i against T gives

$$n_4=8.52867-0.013406T_4 \quad R^2=0.99589 \tag{3.15a}$$

$$n_5=7.80805-0.012178T_5 \quad R^2=0.99625 \tag{3.15b}$$

In order to evaluate the relationship between L_i and temperature T_i, we define $r_i=\partial B_i(T_4)/\partial T$ and plot r_i and L_i against temperature T in Figure 3.1b and 3.1c. Again, strong linearity can be seen. This is because L_i is determined by r_i and $B_i(T_4)$, and both r_i and $B_i(T_4)$ change linearly with temperature. Correlating L_i to T_i, we get

$$L_4=-65.51624683+0.441839238T_4 \quad R^2=0.999275 \tag{3.16a}$$

$$L_5=-70.00300466+0.475706724T_5 \quad R^2=0.999274 \tag{3.16b}$$

Extremely high R^2 indicates that the approximation of L_i as a linear function of temperature is very successful. Thus, for the following derivation, we approximate the parameter L_i as:

$$L_i=a_i+b_iT_i \tag{3.17}$$

Coefficients a_i and b_i are given as follows: a_4=-65.51615 and b_4=0.44184 for channel 4 and a_5=-70.003 and b_5=0.47571 for channel 5. Using the approximation of (3.13) and (3.17), we can derive split window algorithm from (3.10).

3.1.3 Derivation of two-factor split window algorithm for AVHRR data

The derivation of the split window algorithm is based on equation (3.9). For simplification of derivation, we define

$$C_i=\varepsilon_i\tau_i(\theta) \tag{3.18}$$

$$D_i=(1-\tau_i(\theta))[1+(1-\varepsilon_i)\tau_i(\theta)] \tag{3.19}$$

Thus, equation system (3.10) can be rewritten as

$$B_4(T_4)=C_4B_4(T_s)+D_4B_4(T_a) \tag{3.20a}$$

$$B_5(T_5)=C_5B_5(T_s)+D_5B_5(T_a) \tag{3.20b}$$

Applying Taylor expansion equation (3.13) to the equation system, we get

$$L_4\partial B_4(T_4)/\partial T=C_4(L_4+T_s-T_4)\partial B_4(T_4)/\partial T+D_4(L_4+T_a-T_4)\partial B_4(T_4)/\partial T \quad (3.21a)$$

$$(L_5+T_5-T_4)\partial B_5(T_4)/\partial T=C_5(L_5+T_s-T_4)\partial B_5(T_4)/\partial T+D_5(L_5+T_a-T_4)\partial B_5(T_4)/\partial T \quad (3.21b)$$

Elimination of the term $\partial B_4(T_4)/\partial T$ from (3.21a) and $\partial B_5(T_4)/\partial T$ from (3.21b) lets to

$$L_4=C_4(L_4+T_s-T_4)+D_4(L_4+T_a-T_4) \quad (3.22a)$$

$$L_5+T_5-T_4=C_5(L_5+T_s-T_4)+D_5(L_5+T_a-T_4) \quad (3.22b)$$

Elimination of T_a from above equations gives to

$$D_5L_4\text{-}D_4(L_5+T_5-T_4)=D_5C_4(L_4+T_s-T_4)\text{-}D_4C_5(L_5+T_s-T_4)+ D_5D_4(L_4-T_4)-D_4D_5(L_5-T_4) \quad (3.23)$$

Reorganizing equation (3.23) to express T_s in terms of T_4 and T_5, we obtain the general form of split window algorithm as follow

$$T_s=T_4+A(T_4-T_5)+B \quad (3.24)$$

where the coefficients A and B are defined as

$$A=\frac{D_4}{D_5C_4-D_4C_5} \quad (3.25)$$

$$B=\frac{L_4D_5(1-C_4)-L_5D_4(1-C_5)-D_4D_5(L_4-L_5)}{D_5C_4-D_4C_5} \quad (3.26)$$

Because the coefficient B includes the parameter L_i, which is the function of temperature, we need to continue our derivation. For simplification, we donate

$$E_1=\frac{D_5(1-C_4)}{D_5C_4-D_4C_5} \quad (3.27a)$$

$$E_2=\frac{D_4(1-C_5)}{D_5C_4-D_4C_5} \quad (3.27b)$$

$$E_3=\frac{D_4D_5}{D_5C_4-D_4C_5} \quad (3.27c)$$

Substituting into equation (3.26) and replacing L_i with equation (3.17), we get

$$B=E_1(a_4+b_4T_4)-E_2(a_5+b_5T_5)-E_3[(a_4+b_4T_4)-(a_5+b_5T_5)] \quad (3.28)$$

Substituting into equation (3.24) and reorganizing the terms, we get a new form of split window algorithm as follows:

$$T_s=A_0+A_1T_4-A_2T_5 \quad (3.29)$$

where the coefficients A_0, A_1 and A_2 are defined as

$$A_0=E_1a_4-E_2a_5-E_3(a_4-a_5) \quad (3.30a)$$

$$A_1=1+A+(E_1-E_3)b_4 \quad (3.30b)$$

$$A_2=A+(E_2-E_3)b_5 \quad (3.30c)$$

This algorithm relates the determination of the three important coefficients A_0, A_1 and A_2 to the atmospheric transmittance, ground emissivity and viewing angle. The viewing angle is known for specific images. The ground emissivity can be estimated from the combination of daytime and night-time AVHRR data (Li and Becker, 1993). The atmospheric transmittance in the thermal window is generally estimated from water vapour content in the atmospheric profile, which is available in our region. Thus, this algorithm provides a method for our analysis of LST change in the region.

3.2 Determination of atmospheric transmittance

The atmospheric transmittance is a critical parameter that affects the accuracy of LST retrieval using a split window algorithm. The thermal radiance is attenuated on its way to the remote sensor. Transmittance depicts the magnitude of the attenuation of thermal radiance transferring through the atmosphere. It varies with the wavelength and viewing angle. Many atmospheric constituents such as water vapour, O_3, CO_2 and other gases have impacts on the thermal atmospheric transmittance. However, O_3, CO_2 and other gases contents are relatively stable, they can be assumed as a constant and simulated by standard atmospheric profiles. On the contrary, water vapour content is highly variable. Thus, the variation of atmospheric transmittance strongly depends on the dynamics of water vapour content in the profile. Consequently, many split window algorithms relate the determination of atmospheric transmittance to the change of water vapour content while assuming other impacts as constant (Sobrino et al., 1991; França and Cracknell, 1994; Coll et al., 1994a).

3.2.1 Impact of water vapour on atmospheric transmittance

Due to many technical difficulties, the atmospheric transmittance is usually not available *in situ* satellite pass. Generally, the practical way for the determination of atmospheric transmittance is through simulation with local atmospheric conditions, especially water vapour content. Simulation of the relationship between atmospheric transmittance and water vapour content can be done through atmospheric modelling programs such as LOWTRAN and MODTRAN. Here I use the LOWTRAN-7 to determine this relationship. Two atmospheric profiles are used for this simulation: summer and winter. For summer I assume the air temperature near surface being equal to 30°C and for winter 18°C, which represents the case of air temperature change in the study area, but could be extended to the low-to-mid latitude region of the Earth. The viewing angle is also important in the simulation and for this reason I select 10° from nadir because this represents the average zenith angle of the images selected for the analysis.

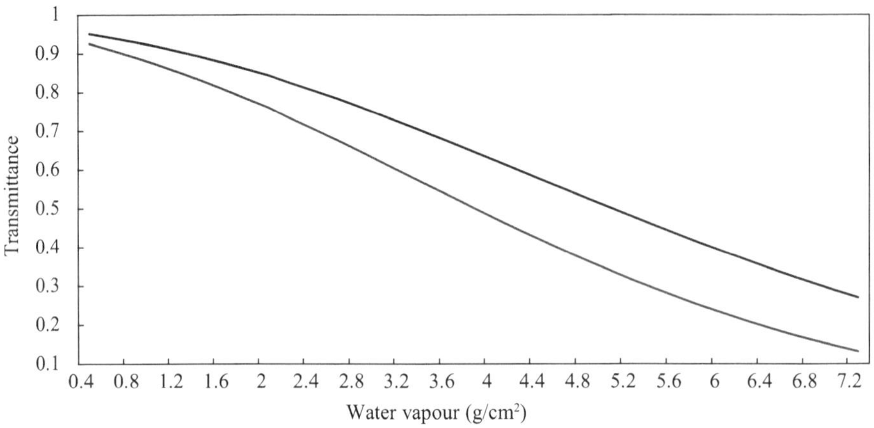

Figure 3.2 Total atmospheric transmittance as a function of water vapour content in summer profile.

Due to arid environments, the water vapour content in the atmospheric column in the region is small. Generally speaking, it is in the range of 1.0-2.2g/cm^2. In order to reveal the change of atmospheric transmittance with water vapour content, I conduct the simulations for the range of 0.4-7.2g/cm^2. Figure 3.2 depicts the results of transmittance simulations for the summer profile. Figure 3.2 indicates that atmospheric transmittance does change with water vapour content in the two AVHRR thermal channels. Atmospheric transmittance in these two channels is high up to about 0.90 for water vapour < 1 g/cm^2 while it is below 0.5 for water vapour >5g/cm^2.

Within a small range of water vapour content, the relationship between water vapour content and atmospheric content can be viewed as linear even though the whole one is a curve as described by França and Cracknell (1994) and Sobrino et al. (1991). In order to have a more accurate estimation of transmittance, I divide the whole curve into several parts and correlate them. Considered the possibility in our case, the results are only given for 0.4-3.0g/cm^2 in Table 3.1. Squared correlation coefficients (R^2) of the transmittance estimation equations given in Table 3.1 is high above 0.995, which indicates that the atmospheric transmittance has strongly linear relation with water content. Standard error illustrates that the estimation of transmittance is with an accuracy of high up to ≤ 0.003 by these equations. This accuracy is much higher than that estimated by a parabolic equation as proposed by França and Cracknell (1994) and Sobrino et al. (1991). Using the parabolic equation to fit the curve, I find that standard error is high up to 0.01015-0.01481 for the range 0.4-6g/cm^2. This error together with possible water vapour content measuring error may cause an obvious error in LST estimation, as illustrated in the next section. The regression coefficients of the equations in Table 3.1 indicate that transmittance change is within a range of 0.006–0.0164 for a small change (0.1g/cm^2) of water vapour content. This means that the maximum of possible error of transmittance estimation is less than 0.033 for a probable water vapour content error 0.2g/cm^2. Usually the satellite *in situ* measurement of atmospheric water vapour content such as CIMEL on the roof of our laboratory can meet this accuracy (water vapour measurement error ≤ 0.2g/cm^2) for transmittance estimation.

Table 3.1 Estimation of atmospheric transmittance for NOAA-AVHRR channels 4 and 5

Profiles	Water vapour (w) (g/cm^2)	Transmittance estimation equation	Squared correlation R^2	Standard error
Summer	0.4-1.6	$\tau_4(10)=0.979160-0.062918w$	0.99425	0.002266
		$\tau_5(10)=0.968144-0.098942w$	0.99716	0.002501
	1.6-3.0	$\tau_4(10)=1.035378-0.097514w$	0.99746	0.002602
		$\tau_5(10)=1.026468-0.135133w$	0.99879	0.002486
Winter	0.4-1.6	$\tau_4(10)=0.983311-0.072444w$	0.99469	0.002215
		$\tau_5(10)=0.981868-0.121979w$	0.99679	0.002896
	1.6-3.0	$\tau_4(10)=1.058059-0.121354w$	0.99817	0.002749
		$\tau_5(10)=1.048364-0.163678w$	0.99948	0.001973

3.2.2 Viewing angle and atmospheric transmittance

The viewing direction of remote sensor determines the pathway of the radiance through the atmosphere to reach the sensor. Zenith angle of viewing (ZAV, also donated as θ) describes the viewing direction of the sensor from the nadir. Even though the possible error of LST retrieval with split window algorithm is negligible at a high viewing angle such as ≥75° (Wan and Dozier, 1996), we still want to analyse the effect of ZAV on atmospheric transmittance. The span of ZAV for the analysis is up to 35° from nadir. Atmospheric water vapour content is selected as to be at the typical level of 2.0g/cm^2 when transmittance is about 0.7-0.85.

Simulation with LOWTRAN program indicates that atmospheric transmittance decreases with ZAV for the two NOAA-AVHRR. This is because the path of radiance transferring through the atmosphere increases with ZAV. Atmospheric transmittance difference from the transmittance with θ=10°, *i.e.* $\tau_i(10)$-$\tau_i(\theta)$, is within 0.04 for ZAV less than 35°. Specifically, for θ=25°, the difference is about 0.02, which may result in an average LST estimation error of about 0.15°C according to sensitivity analysis given in the next section.

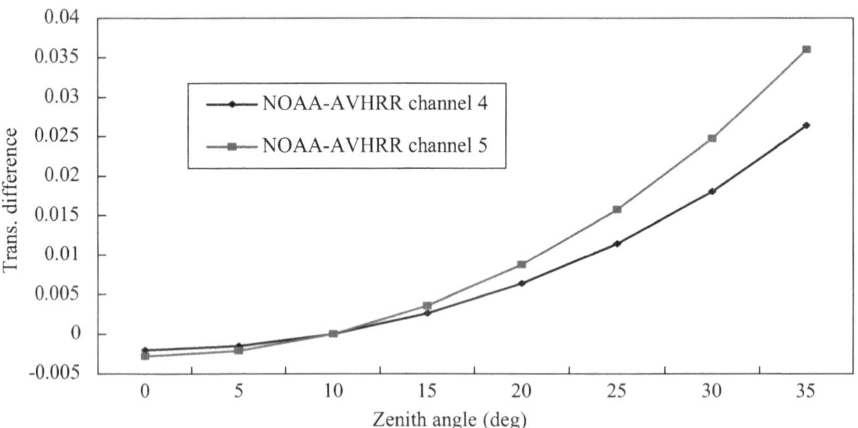

Figure 3.3 Atmospheric transmittance difference $\tau_i(10)$-$\tau_i(\theta)$ as a function of ZAV

Due to the heterogeneous atmospheric conditions in each layer, the relationship between atmospheric transmittance decreasing and ZAV increasing is not linear but parabolic. Figure 3.3 clearly shows that the difference of atmospheric transmittance with θ=10° increases rapidly with ZAV. This feature implies that ZAV has to be considered in the atmospheric transmittance estimation for reaching a relatively accurate LST retrieval from the images with a ZAV of greater than 20°. Using a parabolic function to fit the change of transmittance difference with ZAV, I get the following ZAV correction equation for AVHRR channels 4 and 5:

$$\Delta\tau_4(\theta)=(-2.399387E-3)+(2.29757E-5)\theta^2 \qquad R^2=0.998487 \qquad (3.31a)$$

$$\Delta\tau_5(\theta)=(-3.276602E-3)+(3.14538E-5)\theta^2 \qquad R^2=0.998729 \qquad (3.31b)$$

where $\Delta\tau_4(\theta)$ and $\Delta\tau_5(\theta)$ are atmospheric transmittance difference of AVHRR channels 4 and 5 and TM channel 6, θ is ZAV in degree. Therefore, for an image with ZAV of θ, we can estimate the transmittance in AVHRR channels 4 and 5 as follows:

$$\tau_4(\theta)=\tau_4(10)-\Delta\tau_4(\theta) \qquad (3.32a)$$

$$\tau_5(\theta)=\tau_5(10)-\Delta\tau_5(\theta) \qquad (3.32b)$$

where $\tau_4(10)$ and $\tau_5(10)$ are the atmospheric transmittance of AVHRR channels 4 and 5 at ZAV of 10°, given by the equations listed in Table 3.1.

3.3 Sensitivity analysis of the split window algorithm

Though the estimation of atmospheric transmittance can reach a high accuracy using the above method, there still may be some errors due to the possible effect of other neglected factors such as the variaty of other gases in the atmosphere or/and the accuracy of the water vapour measurement itself. Besides, due to many difficulties, the estimation of ground emissivity always carries some unknown errors (Humes et al., 1994; Becker and Li, 1990). In order to evaluate the impact of these possible errors on LST estimation, it is necessary to perform sensitivity analysis of the split window algorithm. For convenience of expression, the probable LST estimation error δT_s is computed as follows:

$$\delta T_s=|T_s(x+\delta x)-T_s(x)| \qquad (3.33)$$

where x is the variant that sensitivity analysis orients to (transmittance or emissivity), δx is possible error of the variant x, $T_s(x+\delta x)$ and $T_s(x)$ are the LST simulated by our algorithm (3.30) for $x+\delta x$ and x respectively.

3.3.1 Sensitivity analysis to atmospheric transmittance

The sensitivity analysis of the algorithm to transmittance is executed on the basis of several conditions. First, it has to assume a ground emissivity and brightness temperature for the two channels. In the analysis, I use the assumptions of average ground emissivity 0.97 and follow the method created by Sobrino and Caselles (1991) to determine the emissivity of the two channels as $\varepsilon_4=0.967$ and $\varepsilon_5=0.971$. For brightness temperature I assume that T_4 is greater than T_5 and that their difference is $T_4-T_5=0.7$. For most natural surface of the earth, ground emissivity is about 0.95-0.98 (Humes et al., 1994). As shown in Figure 3.4, the $T_4-T_5=0.7$ assumption is rational for most cases. Then, it has to assume a temperature range for the analysis. Considered the possible LST change of the Earth, 0-70°C is selected as the range of brightness temperature T_4 change.

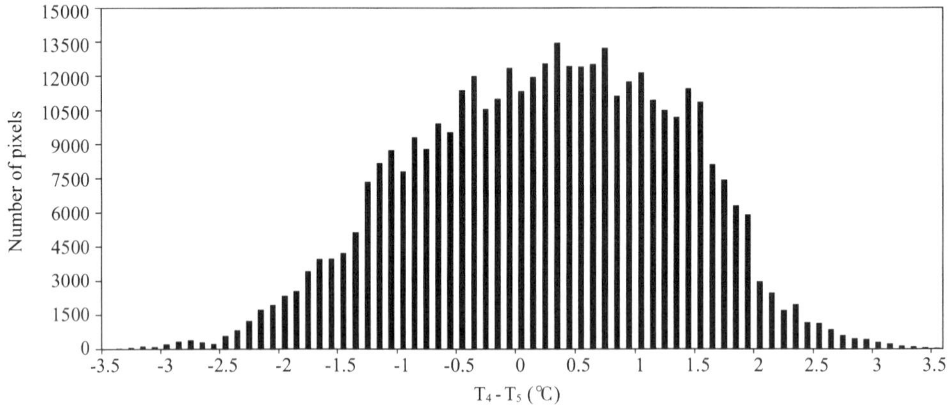

Figure 3.4 Histogram of brightness temperature difference between channels 4 and 5, computing from the land subset of the Israel-Sinai daytime image acquired on July 9, 1998.

The result computed by the algorithm indicates that the LST estimation error is almost independent with temperature change. For $\delta\tau_4$=0.005, the δT_s only changes about 0.0052°C within the temperature range 0-70°C, or from 0.132°C at 0°C to 0.137°C at 70°C. This small change is negligible in practice. Thus, the average δT_s for the range 0-70°C is used to represent the LST estimation error in the following analysis.

Figure 3.5a and 3.5b plot the change of probable δT_s against transmittance due to separate transmittance error in channels 4 and 5, respectively. As shown in Figure 3.5a, the probable LST estimation error is about 0.21-0.361°C for $\delta\tau_4$=0.01 when τ_4 is within the range of 0.805-0.905. The LST error may reach to about 0.393-0.623°C for $\delta\tau_4$=0.02 and 0.695-1.098°C for $\delta\tau_4$=0.05 in the same transmittance range. However, the error is below 0.2°C when channel 4 transmittance is below 0.9 and $\delta\tau_4$ less than 0.005. Similar possible LST estimation error can be seen in the small change of transmittance in channel 5 (Figure 3.5b).

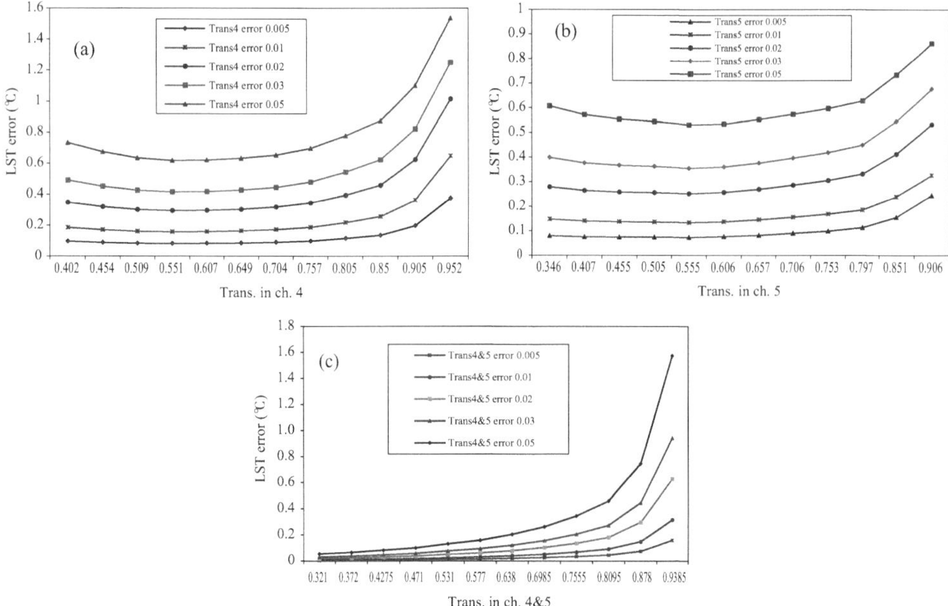

Figure 3.5 Probable LST estimation error with transmittance due to possible transmittance error in (a) channel 4, (b) channel 5 and (c) both channels 4&5.

Because transmittance is estimated by water vapour content in the atmosphere, the separate error of channel 4 or 5 transmittance is almost impossible. Instead, the transmittance error occurs simultaneously on both channels 4 and 5. Figure 3.5c shows the possible LST estimation error due to the possible transmittance error simultaneously occurring in both channels 4 and 5. Comparing Figure 5c with 5a and 5b, one can find that possible error of LST estimation due to the simultaneous transmittance error of channels 4 and 5 is much lower than the separate one (Figure 3.5a and 3.5b). This is because the simultaneous transmittance error of channels 4 and 5 is with accordance to the relative identical change of transmittance in the two channels, which can be seen from the regression coefficients of transmittance estimation listed in Table 3.1. Therefore, for the simultaneous transmittance error $\delta\tau_4=\delta\tau_5=0.01$, the possible LST estimation is about 0.068-0.148°C (Figure 3.5c) when τ_4 is about 0.805-0.905 which is corresponding to τ_5 in the range of about 0.705-0.851 (see Figure 3.2). The error is about 0.136-0.297 for $\delta\tau_4=\delta\tau_5=0.02$ and 0.345-0.748°C for $\delta\tau_4=\delta\tau_5=0.05$ in the same transmittance.

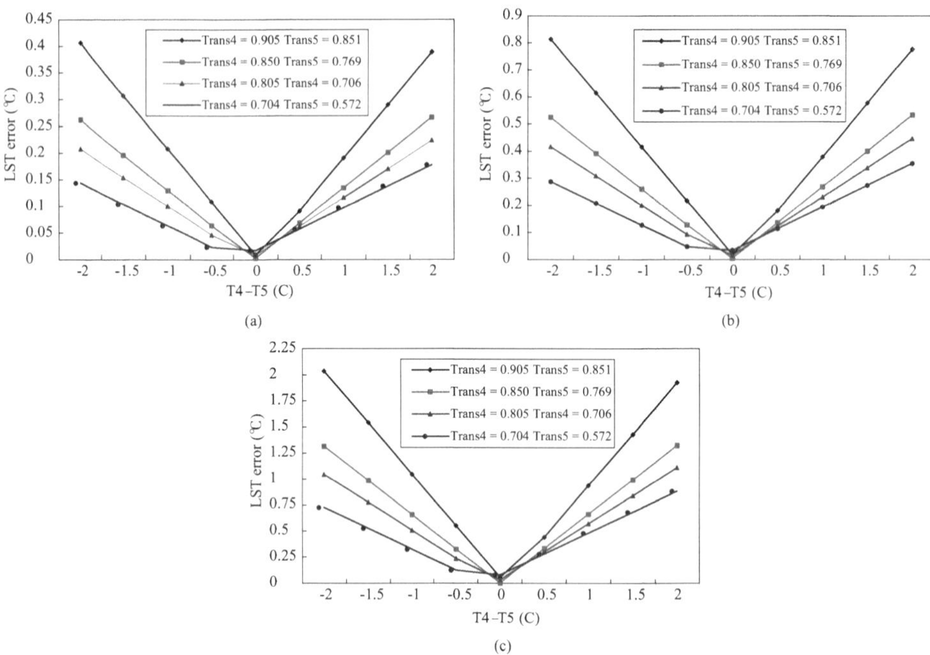

Figure 3.6 Probable LST estimation error with brightness temperature difference due to possible transmittance errors (a) $\delta\tau_4$=$\delta\tau_5$=0.01, (b) $\delta\tau_4$=$\delta\tau_5$=0.02 and (c) $\delta\tau_4$=$\delta\tau_5$=0.05.

The above analysis is based on the assumption of T_4-T_5=0.7. As shown in Figure 3.4, the difference of brightness temperature T_4-T_5 is actually ranging from about -3.5°C to 3.5°C even though above 95% pixels concentrate within the range of ±2°C and above 60% within ±1°C. Therefore, the above analysis based on the assumption T_4-T_5=0.7°C can only represent the average case. In order to evaluate the sensitivity of LST estimation to transmittance, we need to consider a wider range of brightness temperature difference. Figure 3.6 shows the change of possible LST estimation error with T_4-T_5 change. When $|T_4$-$T_5|$ is small, the LST estimation error is small in all cases. The δT_s increases rapidly with the increase of $|T_4$-$T_5|$. Separate transmittance error in channels 4 and 5 causes much higher LST estimation error than simultaneous one. Specifically, for a small error $\delta\tau_4$=$\delta\tau_5$=0.01, possible LST estimation error might be up to above 0.4°C (Figure 3.6a). Higher LST error changes with brightness temperature difference are seen in Figure 3.6b for $\delta\tau_4$=$\delta\tau_5$=0.02 and Figure 3.6c for $\delta\tau_4$=$\delta\tau_5$=0.05. This indicates that the accurate estimation of transmittance such as with an error $\leqslant 0.03$ is the essential prerequisite for an accurate retrieval of LST from remote sensing data. Another implication of the result is that the maximal possible LST estimation error occurs at the pixels with big brightness temperature difference between channels 4 and 5. This conclusion can help us to precisely evaluate the accuracy of LST estimation in a pixel scale when split window algorithm is applied to the real world. When $|T_4\text{-}T_5|$ is small such as less than 0.5°C in the study region, application of the split window algorithm still can produce a very accurate LST estimated in spite of the big transmittance error.

The change of the average LST estimation error with possible transmittance error is shown in Figure 3.7. This average LST error is computed under the conditions $|T_4\text{-}T_5|=1$°C with transmittance range 0.70-0.90 and emissivity range 0.95-0.98. Figure 3.7 indicates that separate transmittance error causes much higher average LST estimation error. For example, an error of 0.05 in τ_4 or τ_5 may result in δT_s of high up to 0.6-0.8C, while the same error in both τ_4 and τ_5 can only produce δT_s of about 0.4°C. Moreover, the relationship between average δT_s and simultaneous transmittance error is almost linear. The average δT_s increases from 0.079K for simultaneous transmittance error $\delta\tau_4=\delta\tau_5=0.01$ through 0.16°C for $\delta\tau_4=\delta\tau_5=0.01$ and 0.23°C for $\delta\tau_4=\delta\tau_5=0.03$ to 0.4°C for $\delta\tau_4=\delta\tau_5=0.05$ (Figure 3.7). Generally, the measurement of water vapour content for transmittance estimation can reach the accuracy of less than 0.2g/cm^2. As mentioned above, the possible estimation error of transmittance due to measurement error of water vapour content in the atmosphere is generally $\leqslant 0.035$.

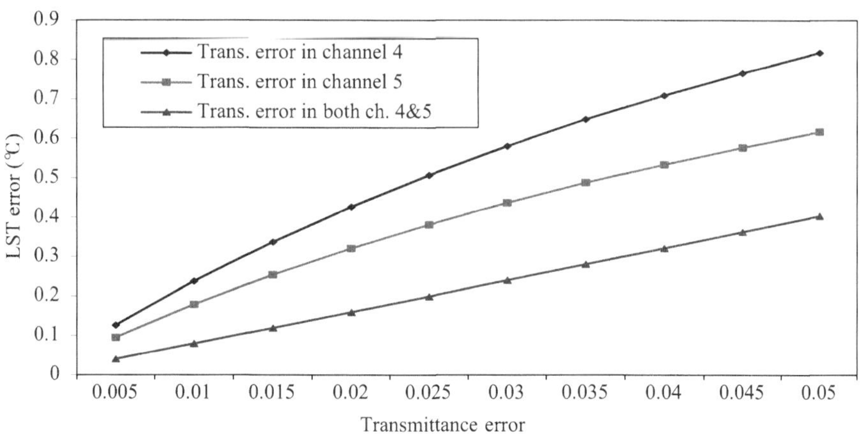

Figure 3.7 Average LST estimation error due to transmittance error.

3.3.2 Sensitivity analysis of ground emissivity

The probable error of LST retrieval due to emissivity error has been discussed in various papers such as Becker (1987), Li and Becker (1993) and Coll et al. (1994a). Based on their algorithm, Li and Becker (1993) estimated the impact of ground emissivity error on probable LST error as a function of average emissivity error $\delta\varepsilon$ and emissivity difference error $\delta(\varepsilon_4-\varepsilon_5)$ between channels 4 and 5. They concluded that the probable LST estimation error may be as large as up to 1.6K for $\delta\varepsilon=\delta(\varepsilon_4-\varepsilon_5)=0.01$. Thus, a good estimation of ground emissivity is the basis for an accurate estimation of LST from remote sensing data.

The sensibility of my newly proposed split window algorithm to ground emissivity variation is shown in Figure 3.8 and 3.9. Several simulations have been done for the sensibility analysis: probable LST estimation error against the change in emissivity and its difference in the two channels as well as in the change of brightness temperature and its difference in the two channels. All the analysis is oriented to possible emissivity error 0.001, 0.002, 0.003 and 0.005. Figure 3.8a shows that probable LST estimation error is not very sensitive to the change in ground emissivity but does depend on the possible emissivity error. For a possible emissivity error of 0.001 in both channels 4 and 5, the change of LST error is less than 0.1°C, ranging from 0.081°C for $\varepsilon_5=0.901$ to 0.065°C for $\varepsilon_5=0.991$. The change of LST error against ground emissivity is linear, with a decrease rate of 0.0156. Moreover, the probable δT_s is also linearly increasing with the possible emissivity error. For an emissivity error of 0.005, the LST error is five times the LST error for an emissivity error of 0.001. Therefore, for an emissivity error 0.10, the probable δT_s is expected to be about 0.65-0.81°C for the emissivity ranging from 0.991to 0.901 (Figure 3.8a).

The LST error is not sensitive to the variation of ground emissivity difference between channels 4 and 5. The probable LST estimation error changes little against the change of emissivity difference $\delta\varepsilon=\varepsilon_4-\varepsilon_5$. For a specific emissivity error such as $\delta\varepsilon_4=\delta\varepsilon_5=0.005$, the possible δT_s changes from 0.36°C at $\delta\varepsilon= -0.01$ through 0.35°C at $\delta\varepsilon=0$ to 0.34°C at $\delta\varepsilon=0.01$. This implies that emissivity difference error has no significant effect on the accuracy of LST error estimation. However, Figure 3.8b does show that the LST error increases steadily with the emissivity error. For an emissivity error 0.001, the δT_s may be 0.07°C and the error increases to 0.35°C for emissivity error 0.005.

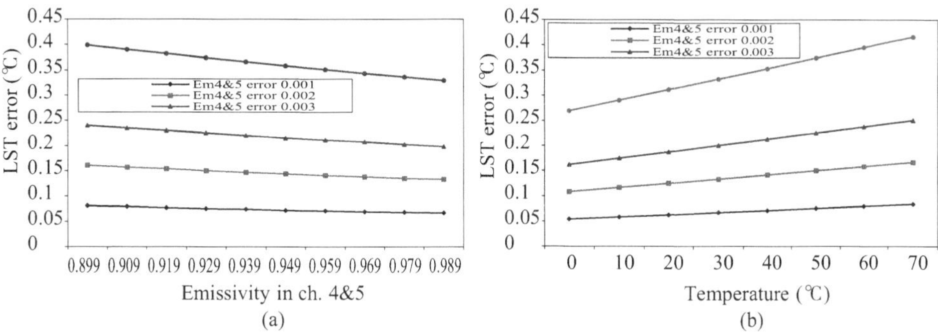

Figure 3.8 Probable LST estimation error due to simultaneous emissivity errors on both channels 4 and 5, illustrating the change of δT_s against (a) average emissivity $(\varepsilon_4+\varepsilon_5)/2$ and (b) brightness temperature.

The probable LST estimation error due to emissivity error is slightly dependent on temperature level (Figure 3.8b). However, the change of δT_s in temperature range 0-70°C is within 0.03°C for every error 0.001 of emissivity. Thus, for an emissivity error 0.005, the range of δT_s is about 0.15°C, changing from 0.27°C at temperature level 0°C to 0.42°C at the level 70°C. Analysis indicates that the probable LST estimation error also has little dependence on the brightness temperature difference T_4-T_5. Within the most possible range of T_4-T_5 change (2°C), the LST error only increases about 0.022K and this increase is quite small.

The impact of ground emissivity error separately in channels 4 and 5 has also been analyzed on the probable LST estimation error. The results show that separate emissivity error may cause an obviously greater LST error than simultaneous one. For an emissivity error 0.005 in channel 4, the possible LST estimation error would be about 0.8°C at emissivity 0.96. Ground emissivity difference between channel 4 and 5 has little effect on the change of LST error for a fixed emissivity error but the brightness temperature difference between the two channels does have significant effect. However, the LST error due to separate emissivity error is always less than 1°C for an emissivity error of up to 0.005. Because estimation of ground emissivity is simultaneously applied to both channels, an emissivity error of 0.005 rarely appears to be only in one channel. Instead, it is possible to appear in both channels. Therefore, it can be concluded that the accuracy of the algorithm to the probable emissivity error is quite high for general purposes of LST estimation.

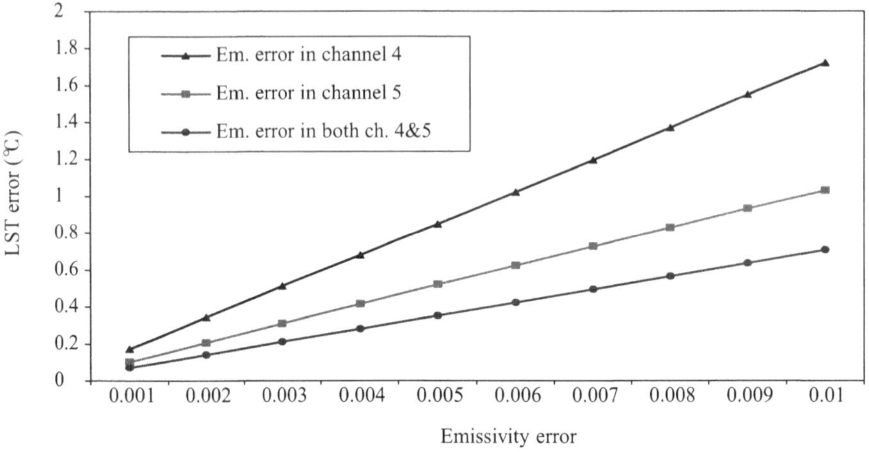

Figure 3.9 Average LST estimation error due to emissivity error.

For a comprehensive evaluation of the sensibility of the algorithm to ground emissivity, the change of average LST estimation error is plotted against emissivity error in Figure 3.9. The condition for computing the average LST estimation error is the same as that of Figure 3.7. Separate emissivity error in channels 4 and 5 causes much higher average LST error than the simultaneous one. If emissivity in channel 4 has an error of 0.01, the possible average δT_s may be up to 1.72°C. The δT_s is only about 1.03°C for $\delta\varepsilon_5$=0.01. As mentioned above, separate error in channel 4 or 5 rarely exists and instead the error usually occurs in a simultaneous way for the two channels. More fortunately, the average LST estimation error is much lower for a simultaneous emissivity error in both channels 4 and 5 than the LST error for a separate one (Figure 3.9). For $\delta\varepsilon_4$=$\delta\varepsilon_5$=0.01, the average δT_s is about 0.708°C. This error is about a half of that estimated by Li and Becker (1993).

3.4 Seasonal LST change in the region

Using the new split window algorithm and the methodology for determining its coefficients, I processed the available AVHRR data of the region for analysis of LST change on both sides. The images with clear sky with good viewing angles (>60°)

were selected for the processing. Since Sede Boker Receiving Station started to acquire NOAA-AVHRR data from the summer of 1995, the images from summer 1995 to 1998 were processed for the analysis. All the selected daytime images were acquired between about 13:00-15:30. Hence the LST retrieved from these daytime images represents the temperature peak of the acquisition date. Night-time images of 1998 were also processed for analysis of the night-time LST change on both sides. All selected night-time images were acquired between 0:00-2:00. Thus, LST retrieved from these night-time images represents the lowest temperature at about midnight of the day.

3.4.1 Daytime LST change and its difference on both sides

The average LST change of the border region in 1995-1996 was plotted in Figure 3.10a, from which we can see the seasonal change of daily maximal temperature change. The climate of the region can be divided into hot dry season and cool wet season. Hot dry season usually ranges from May to October and cool wet season from November to April of next year. LST change retrieved from daytime AVHRR images of 1995 and 1996 clearly illustrates the difference of cool wet and hot dry seasons of the region. LST in cool wet season is obviously lower and in hot dry season much higher. In December of both 1995 and1996, daytime LST was only about 28-30°C and in February and November of 1996 it was about 34-36°C. LST increases to 39-43°C in April of 1996. However, it is high up to 48-56°C in the hot dry season. The maximal LST in each image of 1995-1996 is shown in Figure 3.10b, from which one can see that some pixels have even higher LST than the average one. This implies that LST in summer may be high up to above 56-58°C in some places.

Another feature shown in Figure 3.10a is that the LST on the Israeli side is always higher. A clearer illustration of LST difference on both sides is given in Figure 3.10c, which indicates that the Israeli side is generally about 2.5-3.5°C hotter than the Egyptian side in summer. LST on the Israeli side is also higher in winter though the difference is not as obvious as in summer. Usually the Israeli side is about 0.5-1.5°C hotter in winter.

In order to have an intensive analysis of LST change on both sides, all images with clear sky in 1997-1998 were selected to process. The results are shown in Figure 3.11 and 3.12. The similar features can be seen from Figure 3.11a and 3.12a, which show the average LST change on both sides. During hot dry season, LST at early

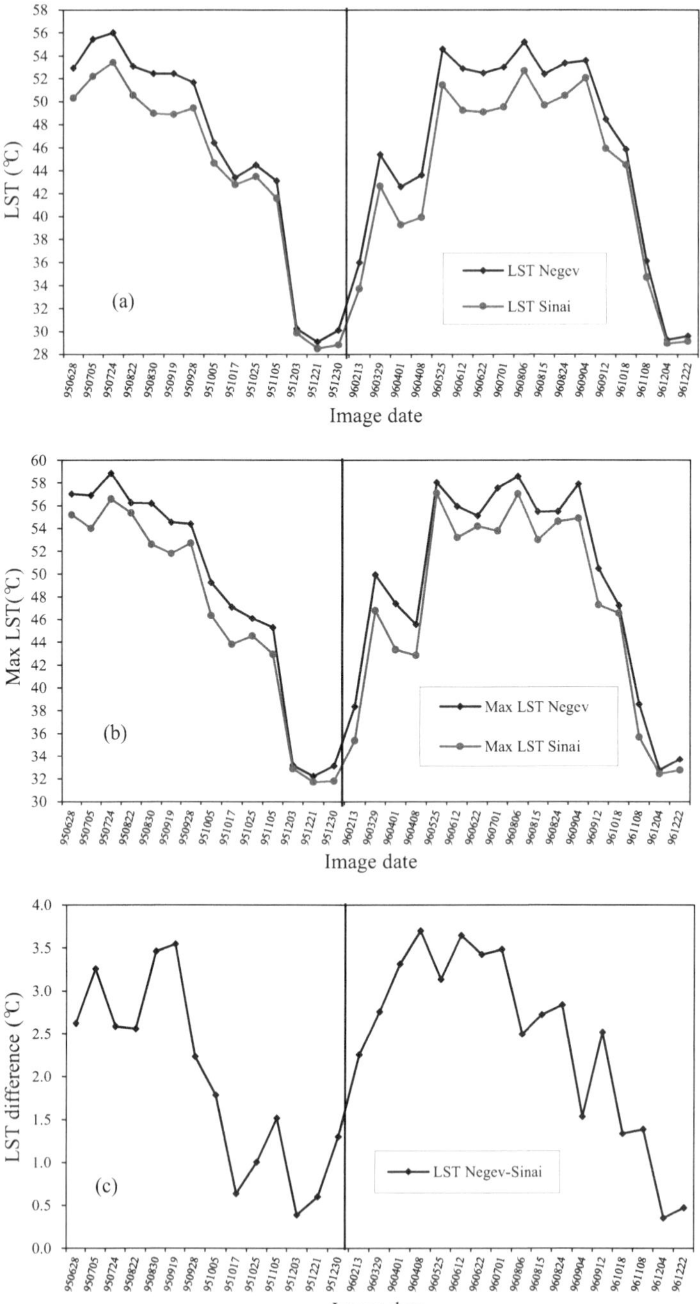

Figure 3.10 LST change on both sides of the region, retrieving from selected NOAA-AVHRR images of 1995-1996, with (a) Average LST change, (b) Maximal LST change, and (c) Average LST difference.

afternoon is very high, fluctuating between 46°C and 56°C. In cool wet season, it is usually lower than 35°C. In the case after heavy rain, the ground surface has several days experiencing extremely low LST (<30°C). Such examples can be seen between January 26 and February 7, 1997 (Figure 3.11a).

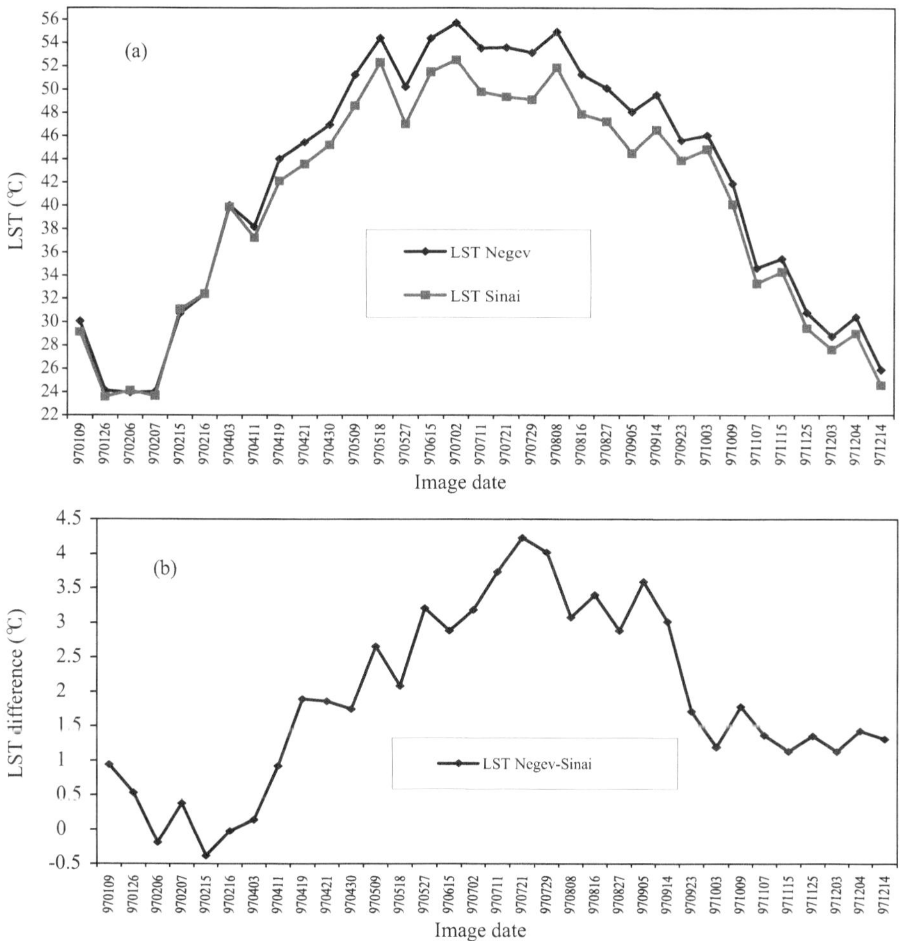

Figure 3.11 LST change on both sides of the region, retrieving from selected NOAA-AVHRR images of 1997, with (a) Average LST change and (b) average LST difference.

As shown in Figure 3.11b and 3.12b, LST difference between Negev and Sinai is the same as that in 1995-1996. In most cases especially in summer, the Israeli side appears to have obviously higher LST. In hot dry season, LST of Negev is about

Figure 3.12 LST change on both sides of the region, retrieving from selected NOAA-AVHRR images of 1998, with (a) Average LST change and (b) average LST difference.

2.5-3.5°C higher. In some cases, the LST difference on both sides can even reach the level of high up to above 4°C. Only in a few extreme cases when the surface was very wet after heavy rain the Egyptian side has slightly higher LST. However, the LST difference in these extreme cases is generally within -0.5°C.

The above analysis of LST change retrieved from NOAA-AVHRR data verifies the supposition of a hotter ground surface on the Israeli side. The anomalous LST difference is very obvious in summer, when the Israeli side is about 2.5-3.5°C hotter. In winter, the LST difference still exists, but reduces to below 1.5°C.

3.4.2 Night-time LST change and its difference on both sides

Night-time images of the region in 1998 were also selected for the retrieval of

LST. The same methodology of LST retrieval from daytime images was applied to process the night-time images. Figure 3.13 shows the night-time average LST change and its difference on both sides. The temporal change of night-time LST is of similar features as the daytime one. During hot dry season, night-time LST is generally higher. From Figure 3.13a one can see that the average LST of the region at about midnight is within the range of about 20-26°C in summer and about 10-15°C in winter.

However, the sharp LST contrast on both sides observed in daytime images does not appear in night-time images. As shown in Figure 3.13b, the night-time LST

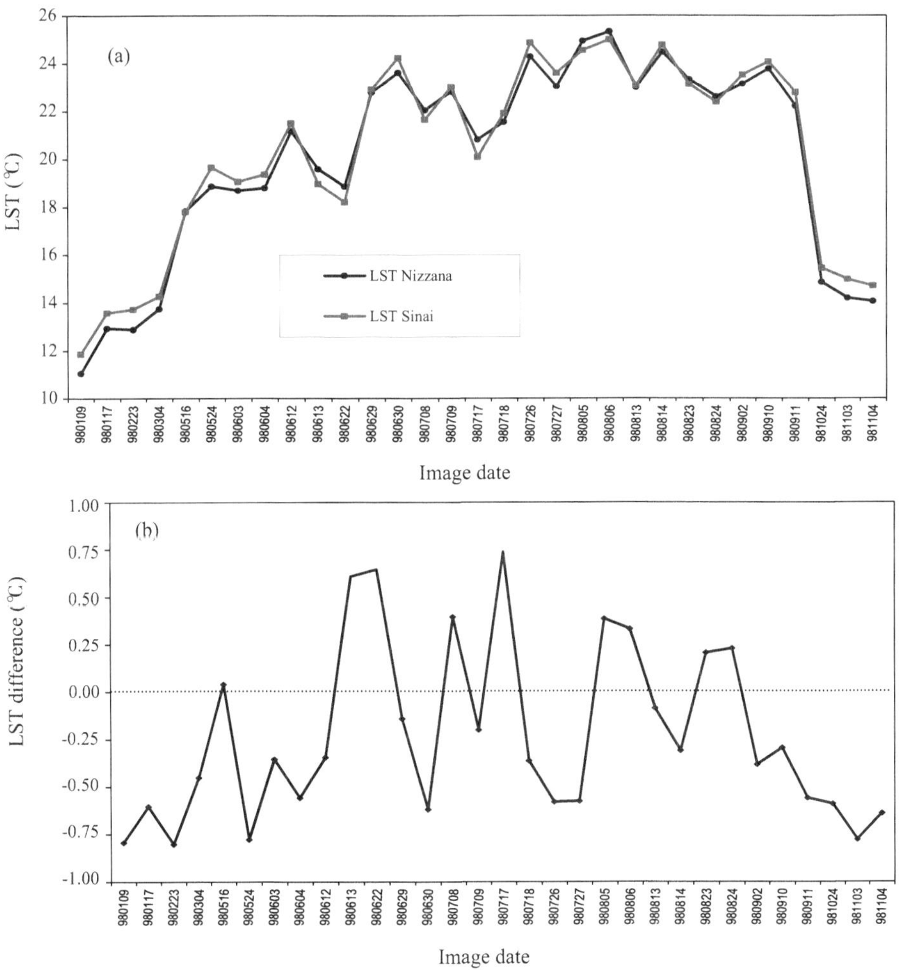

Figure 3.13 Night LST change on both sides of the region, retrieved from selected NOAA-AVHRR night images of 1998. (a) Average LST change, (b) average LST difference.

difference is generally within ±0.75°C, which is quite small compared with the sharp contrast in daytime. Moreover, the difference is not regular. Sometime the Israeli side has higher LST while sometime the opposite change can be seen. Therefore, the conclusion can be drawn that in summer the obvious LST difference during daytime disappears during night-time on both sides. However, in winter it seems that the Egyptian side is a little bit hotter even though the average night-time LST difference is only about 0.5-0.8°C on both sides.

4 LST change from Landsat TM data in the region

The Landsat Thematic Mapper (TM) has a thermal band (channel 6) operating in the wavelength range of 10.4-12.5μm with a ground resolution of 120 m. This spatial resolution is much higher than the nadir ground resolution (1.1km) of AVHHR. Therefore, it is very suitable for use to analyze the detailed spatial patterns of LST distribution and its difference on both sides of the region. Retrieval from LST from Landsat TM6 data can be divided into two steps: calculating the brightness temperature from the radiance and then considering the effects of atmospheric transmittance and ground emissivity for LST estimation.

The Landsat Thematic Mapper (TM) images have been extensively studied for various purposes (Caselles et al., 1998; Lo, 1997; Kaneko and Hino, 1996). Except the 6 bands in visible and near infrared (NIR) wavelengths, the remote sensor also has a thermal band (TM6) operating in the wavelength range of 10.45-12.50 μm with a nominal ground resolution of 120m×120m. This spatial resolution of Landsat TM6 is high enough for analyzing the detailed spatial patterns of thermal variation on the Earth's surface. However, the study of Landsat TM thermal band for surface temperature and its application still remains as an ignored area. This is especially true when referring to land surface temperature (LST). Due to its high spatial resolution, Landsat TM6 has considerable potential for many applications relating to the LST. Some studies relating to the thermal band of Landsat TM only use the brightness temperature at the satellite level (Zhang et al., 1997; Saraf et al., 1995; Mansor et al., 1994) or just simply use the digital number (DN) value for their applications (Oppenheimer, 1997; Ritchie et al., 1990). The actual use of LST retrieved from Landsat TM6 is few (Sospedra et al.,1998, Hurtado et al., 1996). In addition to its intrinsic weaknesses (no on-board calibration, low repeat frequency, and so on), the lack of a proper and easily-used algorithm for retrieval of LST from the only one thermal band of Landsat TM data probably is also the main reason leading to fewer application.

Up to present, the studies of Landsat TM thermal data mainly concentrate on the level of directly applying brightness temperature or DN value to the issues in the real world and on the study of the sea/lake surface temperature corrected with atmospheric model using radiosonde data. Zhang et al. (1997), Saraf et al. (1995) and Mansor et al. (1994) demonstrated that the radiant temperature converting from Landsat TM thermal data is capable of application for detection of subsurface coal fires. Sugita and Brutsaert (1993) compared the measured LST in the field with the temperature from satellite including Landsat TM. The measured sea surface temperature is found to be about 7.8°C higher than the brightness temperature for Landsat TM6 around the out-fall of a nuclear power plant (Liu and Kuo, 1994). Haakstad et al. (1994) demonstrated the importance of Landsat TM6 data in sea temperature study for identifying the surface current patterns. The relationship between water quality indicators and Landsat TM digital data, including TM6 DN value has been analyzed in the study of Braga et al. (1993) about water quality assessment at Guanabara Bay of Brazil. Oppenheimer (1997) used the relationship between the measured lake surface temperature and Landsat TM6 DN value to map the thermal variation of volcano lakes. Correlation of the lake suspended sediments with digital data of Landsat MSS and TM was analyzed in Rictchie et al. (1990). Based on the surface temperature derived from Landsat TM thermal data, Moran et al. (1989) estimated the latent heat and net radiant flux density and compared with the ground estimate based on Bowen ratio measurement over mature fields of cotton, wheat and alfalfa. The Landsat TM6 data has been used in several studies to investigate the thermal properties of volcanoes (Kaneko, 1996; Andres and Rose, 1995; Reddy et al., 1990)

Provided that the ground emissivity is known (the determination of emissivity is very complicated and there is a great volume of literature on it), the retrieval of LST from Landsat TM6 is mainly through the method of atmospheric correction. The principle of the correction is to subtract the upward atmospheric thermal radiance and the reflected atmospheric radiance from the observed radiance at satellite level so that the brightness temperature at ground level can be directly computed. The atmospheric thermal radiance can be simulated using such atmospheric simulation programs as LOWTRAN, MODTRAN or 6S when in situ atmospheric profile is available at the satellite pass. Usually this is not the case for many applications. Thus, an alternative is to use the available radiosonde data closed to the satellite pass or with similar atmospheric conditions for this atmospheric simulation (Hurtado et al., 1996). In many cases, even the radiosonde data is also not available due to difficulties for the measuring. This unavailability of in situ atmospheric profile data prevents the popular application of LST retrieval from Landsat TM6 for many studies. Alternately, the standard atmospheric profiles provided in the atmospheric simulation programs were used to simulate the atmospheric radiance for retrieval of surface temperature from the Landsat TM thermal data.

4.1 Computing brightness temperature of Landsat TM6 data

The gray level of Landsat TM data is given as digital number (DN) ranging from 0 to 255. Thus, the computation of brightness temperature from TM data includes the estimation of radiance from its DN value and the conversion of the radiance into brightness temperature.

The following equation developed by the National Aeronautics and Space Administration (NASA) (Markham and Barker, 1986) is generally used to compute the spectral radiance from DN value of TM data:

$$L_{(\lambda)} = L_{\min(\lambda)} + (L_{\max(\lambda)} - L_{\min(\lambda)}) Q_{dn}/Q_{\max} \tag{4.1}$$

where $L_{(\lambda)}$ is the spectral radiance received by the sensor (mW $cm^{-2}sr^{-1}\mu m^{-1}$), Q_{max} is the maximal DN value with Q_{max}=255, and Q_{dn} is the gray level for the analyzed pixel of TM image, $L_{\min(\lambda)}$ and $L_{\max(\lambda)}$ are the minimal and maximal spectral radiance for Q_{dn}=0 and Q_{dn}=255, respectively. In fact, it has been set that $L_{\min(\lambda)}$ =0.1238 mW cm^{-2} $ster^{-1}\mu m^{-1}$ for Q_{dn}=0 and $L_{\max(\lambda)}$=1.56 mW cm^{-2} $ster^{-1}\mu m^{-1}$ for Q_{dn}=255. Thus, the above equation can be simplified as the following form (Schneider and Mauser, 1996):

$$L_{(\lambda)}=0.1238+0.005632156Q_{dn} \tag{4.2}$$

Once the spectral radiance $L_{(\lambda)}$ is computed, the brightness temperature at the satellite level can be directly computed with the following equation:

$$T_6=K_2/\ln(1+K_1/L_{(\lambda)}) \tag{4.3}$$

where T_6 is the effective at-satellite brightness temperature of TM6 in K, K_1 and K_2 are pre-launch calibration constants. For Landsat 5, which we will use in the study, K_1= 60.776 mW $cm^{-2}sr^{-1}\mu m^{-1}$ and K_2=1260.56K, respectively.

4.2 Mono-window algorithm for LST retrieval

It has been well known that the impacts of the atmosphere and the emitted ground are unavoidably involved in the sensor-observed radiance. Thus, correction is necessary for retrieving true LST from Landsat TM6 data (Hurtado et al., 1996). The correction is based on the radiance transfer equation, which states that the sensor-observed radiance is impacted by the atmosphere and the emitted ground. Here we considered both the atmospheric and ground emissiviy effects into the thermal radiance transfer equation and developed a mono-window algorithm for LST from the Landsat TM data.

4.2.1 Thermal radiance transfer equation

In accordance with blackbody theory, the thermal emittance from an object can be expressed as Planck's radiance function. Blackbody is only a theoretical concept. Most natural surface is in fact not blackbody. Thus, emissivity has to be considered for constructing the radiance transfer equation. Moreover, while transferring from the emitted ground to the remote sensor, the ground emittance is attenuated by the atmospheric absorption. On the other hand, the atmosphere also contributes the emittance that reaches the sensor either directly or indirectly (reflected by the surface). Considering all these components and effects, the sensor-observed radiance for Landsat TM6 can be expressed as:

$$B_6(T_6)=\tau_6[\varepsilon_6 B_6(T_s)+(1-\varepsilon_6)I_6^{\downarrow}]+I_6^{\uparrow} \tag{4.4}$$

where T_s is land surface temperature, and T_6 is brightness temperature of TM6, τ_6 is atmospheric transmittance and ε_6 is ground emissivity. $B_6(T_6)$ is radiance received by the sensor, $B_6(T_s)$ is ground radiance, $I_6^{\downarrow}$and $I_6^{\uparrow}$ are the down-welling and upwelling atmospheric radiances, respectively.

The upwelling atmospheric radiance $I6^{\uparrow}$ is usually computed (França and Cracknell, 1994; Cracknell, 1997) as

$$I_6^{\uparrow}=\int_0^Z B_6(T_z)\frac{\partial\tau_6(z,Z)}{\partial z}dz \tag{4.5}$$

where T_z is atmospheric temperature at altitude z, Z is altitude of the sensor, τ6(z, Z) represents the upwelling atmospheric transmittance from altitude z to the sensor height Z. Following McMillin (1975), Prata (1993) and Coll et al. (1994), we employ the mean value theorem to express the upwelling atmospheric radiance as:

$$B_6(T_a) = \frac{1}{1-\tau_6}\int_0^Z B_6(T_z)\frac{\partial \tau_6(z,Z)}{\partial z}dz \tag{4.6}$$

where T_a is the effective mean atmospheric temperature and $B_6(T_a)$ represents the effective mean atmospheric radiance with T_a for TM6. Thus, we get

$$I_6^{\uparrow}=(1-\tau_6)B_6(T_a) \tag{4.7}$$

The down-welling atmospheric radiance is generally viewed as from a hemispherical direction, hence can be computed (França and Cracknell, 1994) as

$$I_6^{\downarrow} = 2\int_0^{\pi/2}\int_{\infty}^{0} B_6(T_z)\frac{\partial \tau'_6(\theta',z,0)}{\partial z}\cos\theta'\sin\theta'\,dz\,d\theta' \tag{4.8}$$

where θ' is the down-welling direction of atmospheric radiance and $\tau'_6(\theta',z,0)$ represents the down-welling atmospheric transmittance from altitude z to the ground surface. According to França and Cracknell (1994), it is rational to assume that $\partial\tau'_6(\theta',z,0)=\partial\tau_6(z,Z)$ for the thin layers of the whole atmosphere when the sky is clear. Based on this assumption, application of mean value theorem to Equation (4.5) gives

$$I_6^{\downarrow} = 2\int_0^{\pi/2}(1-\tau_6)B_6(T_a^{\downarrow})\cos\theta'\sin\theta'\,d\theta' \tag{4.9}$$

where $T_a^{\downarrow}$ is the downward effective mean atmospheric temperature. The integration term of this equation can be solved as

$$2\int_0^{\pi/2}\cos\theta'\sin\theta'\,d\theta' = (\sin\theta')^2\Big|_0^{\pi/2} = 1 \tag{4.10}$$

Thus, the downward atmospheric radiance cab be estimated as

$$I_6^{\downarrow}=(1-\tau_6)\,B_6(T_a^{\downarrow}) \tag{4.11}$$

Substitution into equation (5) gives to

$$B_6(T_6)=\varepsilon_6\tau_6B_6(T_s)+\tau_6(1-\varepsilon_6)(1-\tau_6)\,B_6(T_a^{\downarrow})+(1-\tau_6)\,B_6(T_a) \tag{4.12}$$

The Solution of the above equation for Ts can give an algorithm for LST retrieval from the Landsat TM data.

4.2.2 Derivation of mono-window algorithm

In order to solve this equation for LST, we need to analysis the effect of $B_6(T_a^{\downarrow})$ on the observed LST by TM6. Due to the vertical difference of atmosphere, the upward atmospheric radiance is generally greater than the downward one. Consequently, $B_6(T_a)$ is greater than $B_6(T_a^{\downarrow})$, or $T_a > T_a^{\downarrow}$. Under the clear sky, the difference between T_a and $T_a^{\downarrow}$

is usually within 5°C, i.e. | T_a-$T_a^{\downarrow}$|<5°C.

For convenience of analysis, we denote $D' = \tau_6(1-\varepsilon_6)(1-\tau_6)$. Since the emissivity ε_6 is generally 0.96-0.98 for most natural surfaces, the value of D' is very small, mainly depended on τ_6. Provided $\tau_6=0.7$ and $\varepsilon_6=0.96$, we get $D'=0.0084$. The very small value of D' makes the approximate of $B_6(T_a^{\downarrow})$ with $B_6(T_a)$ possible and feasible for the derivation of an algorithm from the above equation.

Before we go on our derivation, we need to simulate the effect of approximating $B_6(T_a^{\downarrow})$ with $B_6(T_a)$ on the change of T_s in Equation (13). Since $B_6(T_a)> B_6(T_a^{\downarrow})$, the approximation of $B_6(T_a^{\downarrow})$ with $B_6(T_a)$ would lead to the underestimate of both $B_6(T_s)$ and $B_6(T_a)$ in Equation (13) for the fixed $B_6(T_6)$. Consequently, it will lead to the underestimate of T_s. The magnitude of the underestimate of $B_6(T_s)$ and $B_6(T_a)$ in Equation (13) depends on their coefficients in the equation, i.e. ε6τ6 for $B_6(T_s)$ and $(1-\tau_6)[1+\tau_6(1-\varepsilon_6)]$ for $B_6(T_a)$. We consider three cases of | T_a-$T_a^{\downarrow}$| and two cases of ε6 and τ6 for the simulation, which gives the results shown in Table 1. The underestimates of T_s for all cases are quite small. For |T_a-$T_a^{\downarrow}$|=5°C and τ6=0.8, the approximation of $B_6(T_a^{\downarrow})$ with $B_6(T_a)$ can only lead to the underestimate of T_s 0.0255°C at T_s =20°C and 0.0205°C at T_s =50°C. The underestimates of Ts are even smaller for $\tau_6=0.7$ (Table 1). Therefore, we can conclude that the approximation of $B_6(T_a^{\downarrow})$ with $B_6(T_a)$ will have an insignificant effect on the estimate of T_s from the above equation. With this approximation, the observed radiance of Landsat TM6 can be expressed as

$$B_6(T_6)=\varepsilon_6\tau_6 B_6(T_s)+(1-\tau_6)[1+\tau_6(1-\varepsilon_6)]\,B_6(T_a) \tag{4.13}$$

which enables us to solve T_s for LST retrieval.

Table 4. 1 Underestimate of T_s by approximating $B_6(T_a^{\downarrow})$ with $B_6(T_a)$

$T_a^{\downarrow}$-T_a in °C	Underestimate of T_s in °C					
	For τ_6=0.7 and ε_6=0.96			For τ_6=0.8 and ε_6=0.96		
	At T_s=20	At T_s=35	At T_s=50	At T_s=20	At T_s=35	At T_s=50
2	0.0087	0.0077	0.0070	0.0100	0.0089	0.0081
3	0.0130	0.0116	0.0105	0.0150	0.0134	0.0121
5	0.0220	0.0196	0.0177	0.0255	0.0227	0.0205

In order to solve T_s from equation (4.10), we need to linearize Planck's radiance function. Because the change of Planck's radiance with temperature is very close to linearity in a narrow temperature range (say, <15°C) for a specific wavelength, the linearization of Planck's function can be done by Taylor's expansion for only keeping the first two terms:

$$B_6(T_j)=B_6(T)+(T_j-T)\,\partial B_6(T)/\partial T=(L6+T_j-T)\,\partial B_6(T)/\partial T \tag{4.14}$$

where T_j refers to the brightness temperatures (j=6), land surface temperature (j=s) and mean atmospheric temperature (j=a). The parameter L_6 is defined as

$$L_6=B_6(T)/[\partial B_6(T)/\partial T] \tag{4.15}$$

in which L6 has the dimension of temperature in K. The physical meaning of Taylor's expansion in this case is to expresses the radiance of $B_6(T_j)$ in terms of the radiance $B_6(T)$ with a fixed temperature T. Figure 4.1 shows the change of parameter L6 with temperature for Landsat TM6.

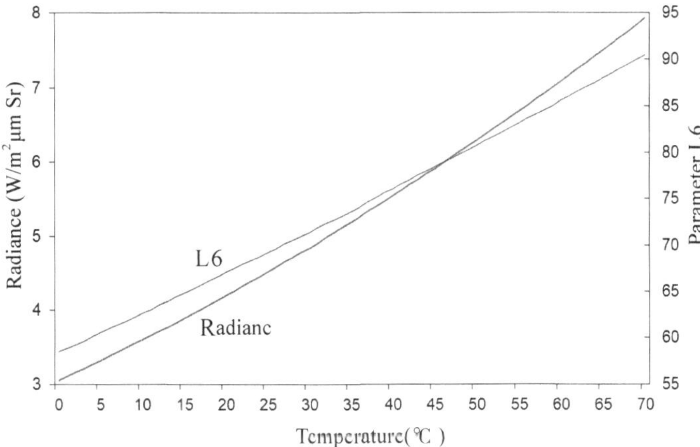

Figure 4.1 Change of Planck function $B_6(T)$ and parameter L_6 with temperature for Landsat TM6.

Considering the possible T_s>T_6>T_a for most cases, we defined T in Taylor's expansion as T_6. Thus, to express the Planck's radiance of Ts and Ta for T6, we have

$$B_6(T_s)=(L_6+T_s-T_6)\,\partial B_6(T)/\partial T \tag{4.16a}$$

$$B_6(T_a)=(L_6+T_a-T_6)\,\partial B_6(T)/\partial T \tag{4.16b}$$

$$B_6(T_6)=(L_6+T_6-T_6)\,\partial B_6(T)/\partial T)=L_6\partial B_6(T)/\partial T \tag{4.16c}$$

Substituting into equation (4.11) and eliminating the term ∂B6(T6)/∂T, we obtain

$$L_6=\varepsilon_6\tau_6(L_6+T_s-T_6)+(1-\tau_6)[1+(1-\varepsilon_6)\tau_6](L_6+T_a-T_6) \tag{4.17}$$

For simplification, we define

$$C_6=\varepsilon_6\tau_6(\theta) \tag{4.18}$$

$$D_6=(1-\tau_6(\theta))[1+(1-\varepsilon_6)\tau_6(\theta)] \tag{4.19}$$

Thus, we have

$$L_6=C_6(L_6+T_s-T_6)+D_6(L_6+T_a-T_6) \tag{4.20}$$

where the parameter L_6 is defined the same as in equation (4.12). For TM6, we also find that L_6 has linear relation with the temperature. Thus we can use following equation to approximate it for the derivation.

$$L_6=a_6+b_6T_6 \tag{4.21}$$

For possible temperature range 10-45°C in the study region, the coefficients of equation (4.18) are computed as a_6=-64.81616 and b_6=0.449637.

Table 4.2 Coefficients of parameter L6 for different temperature ranges

LST Range °C	a_6	b_6	REE %	R^2	T_{test}	F_{test}
0-30	-60.3263	0.43436	0.0833	0.9998	186.7	150147.8
10-40	-63.1885	0.44411	0.0973	0.9997	151.6	100984.8
20-50	-67.9542	0.45987	0.1225	0.9995	117.5	60141.8
30-60	-71.9992	0.47271	0.0621	0.9999	223.9	218819.8

With this relation, we get

$$a_6+b_6T_6==C_6(a_6+b_6T_6+T_s-T_6)+D_6(a_6+b_6T_6+T_a-T_6) \tag{4.22}$$

Solving for T_s, we have the algorithm for LST retrieval from Landsat TM6 data.

$$T_s=[a_6(1-C_6-D_6)+(b_6(1-C_6-D_6)+C_6+D_6)T_6-D_6T_a]/C_6 \tag{4.23}$$

Provided that ground emissivity and atmospheric transmittance are known, the computation of LST from TM6 data is depended on the determination of effective mean atmospheric temperature T_a. Because this algorithm only requires one thermal channel for LST estimation, we define it as mono-window algorithm in order to distinguish from the above split window algorithm.

4.3 Determination of land surface emissivity

Land surface emissivity mainly subjects to two factors: composition of materials on the surface and wavelength of remote sensor. For Landsat TM6, the wavelength is fixed. Thus, the variation of LSE is mainly a function of surface materials. Though the structural composition of surface materials is rather complicated, three categories can be identified in the pixel scale of Landsat TM6: water surface, man-made surface and natural surface. The man-made surface mainly involves cities, towns and villages which can be further distinguished into roads, buildings and some vegetated areas such as parks and gardens inside the cities. The pixels of man-made surface are generally not so many in comparison to those of natural surface, which can be termed as forests, farmlands, deserts and so on. For LST estimation, natural surface usually occupies the most areas of TM images under study, hence should be our focus of analysis.

Actually the TM pixels of natural surface can be simply viewed as composed of vegetation and bare soil in different fractions. Such pixels can be termed as mixed pixels in image processing. The fraction of vegetation cover presents an obvious feature of seasonal vibration due to different canopy density. In this case, we can estimate the total thermal emission from a mixed surface as follows:

$$B(T_s)\varepsilon = P_v B(T_v)\varepsilon_v + (1 - P_v) B(T_{bs})\varepsilon_s + \mathrm{d}\Delta \tag{4.24}$$

where $B(T_s)$ represents the thermal emission of ground surface under its average surface temperature of T_s; $B(T_v)$ and $B(T_{bs})$ are the thermal emissions from vegetated part and bare soil part, respectively; P_v is vegetation fraction of the surface, with $0 \leqslant P_v \leqslant 1$; $\mathrm{d}\Delta$ represents the interaction of thermal emission between vegetation part and bare soil part; ε_v and ε_s are emissivity of vegetation and bare soil, and ε the surface's average emissivity that we intend to solve for TM6 wavelength.

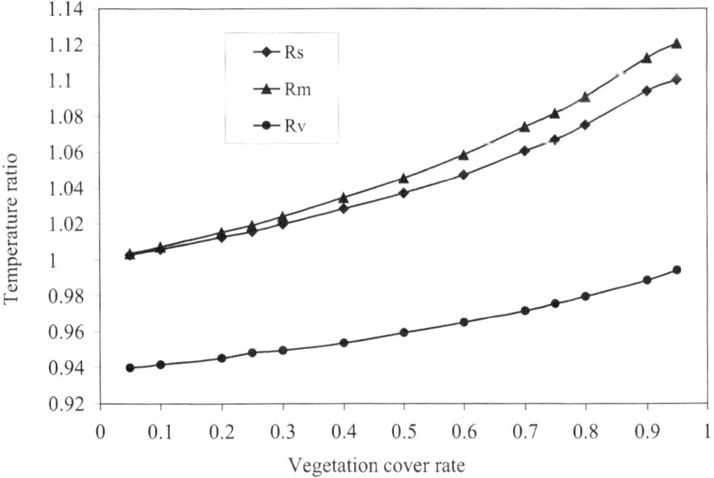

Figure 4.2 Change of radiance ratios R_v, R_s, and R_m for three surface patterns: vegetation, bare soil, and man-made.

As we can imagine, if a pixel of Landsat TM6 images is completely covered with vegetation, the LSE will be the emissivity of vegetation, i.e. $\varepsilon=\varepsilon_v$ if $P_v=1$. Another extreme is bare soil, in which we have $\varepsilon=\varepsilon_s$ for $P_v=0$. However, the natural surface in reality is rather complicated that we rarely have the pure pixels. Therefore, we can use $B(T_s)$ to divide each term on both sides of equation (3.33), which gives

$$\varepsilon=P_vR_v\varepsilon_v+(1-P_v)R_s\varepsilon_s+d\varepsilon \qquad (4.25)$$

This is the equation for LSE estimation, in which R_v and R_s are radiance ratios of vegetation and bare soil, defined as follows for the TM6 wavelength:

$$R_v=B_6(T_v)/B_6(T_s) \qquad (4.26a)$$

$$R_s=B_6(T_{bs})/B_6(T_s) \qquad (4.26b)$$

Similarly, we have the LSE estimation equation for man-made surface which is mixed with vegetation and various buildings as follows:

$$\varepsilon=P_vR_v\varepsilon_v+(1-P_v)R_m\varepsilon_m+d\varepsilon \qquad (4.27)$$

where R_m is the radiance ratio of buildings. Generally speaking, the interaction term $d\varepsilon$ of equation (4.24) or (4.26) is quite small. It can be neglected (i.e. $d\varepsilon=0$) for the regions with flat surface such as plain. For the rough surfaces, the term $d\varepsilon$ can be estimated through vegetation fraction. It is expected that the thermal interaction between vegetation and bare soil may reach maximal in the case when their fraction is equal to 0.5, i.e. each of them occupies half of the ground surface. Though $d\varepsilon$ may vary with heights of vegetation canopy from bare soil surface, it can be roughly estimated as $d\varepsilon=0.001898$ for $P_v=0.5$ according to the studies of Becker and Li (1995) and Humes et al. (1994). In other cases where $P_v<0.5$ or $P_v>0.5$, we have

$$\text{For } P_v<0.5, \quad d\varepsilon=0.003796P_v \qquad (4.28a)$$

$$\text{For } P_v>0.5, \quad d\varepsilon=0.003796(1-P_v) \qquad (4.28b)$$

Therefore, estimation of LSE with equation (4.24) or (4.26) will completely rely on the determination of following variables: radiance ratios R_v and R_b, vegetation canopy emissivity ε_v, bare soil emissivity ε_s, and vegetation fraction P_v. Now let's discuss the determination of radiance ratios first. Since the ratios are temperature dependents, simulation has to be done using Planck function for the wavelength of TM6. The results show that the ratios vary slightly with temperature but increase significantly with vegetation cover percent, as seen in Figure 1. Therefore, we can propose the following regression equations to estimate the magnitude of the ratios:

$$R_v=0.9332+0.0585P_v \tag{4.29a}$$

$$R_s=0.9902+0.1068P_v \tag{4.29b}$$

$$R_m=0.9886+0.1287P_v \tag{4.29c}$$

Our experimental computation indicates that estimation of the ratios can meet the accuracy requirement of LST retrieval even though their slight variations with temperature do not involve. After solving the problem of determining the radiance ratios, we go into the next step of determining vegetation emissivity and soil emissivity, which can be done through examination of thermal spectra of typical materials from JPL spectra da Table ase. Figure 4.3 shows the change of spectral emissivity of typical soils within the thermal ranges. According to the spectral curves, we have the average emissivity of these soils for TM6 wavelength as follows: 0.96866 for brown loam sandy soil, 0.97953 for silt-clay soil, 0.97047 for sandy soil and 0.96993 for sandy loam soil. Even though the soils show some differences in their emissivities, their average value i.e. $\varepsilon_s=0.97215$ can be used for LST retrieval if there is no information about the soils' emissivity in the region under study.

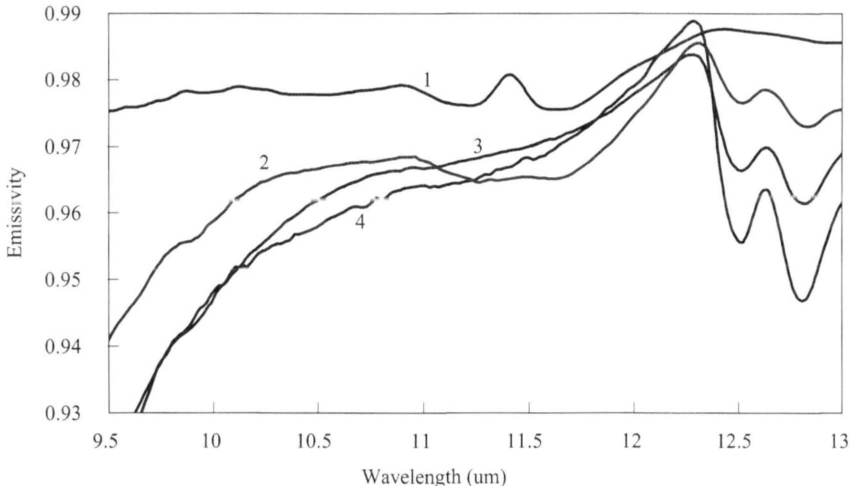

Figure 4.3 Emissivity change of typical soils with wavelength, in which curve 1 represents brown loam sandy soil, 2 silt-clay soil, 3 sandy soil, and 4 sandy loam soil.

As shown in Figure 4.4, emissivity of water and vegetation types is very high. Average emissivity within TM6 wavelength is 0.99499 for fresh water and 0.9951 for sea water. Slight differences of emissivity can be seen among deciduous trees, conifer plants and grass. Vegetation emissivity within TM6 wavelength is 0.9843 for deciduous trees, 0.98803 for conifer plants and 0.9859 for grass. If the vegetation types are not known in study region, one may just use the average emissivity of ε_v=0.98607 for TM6 wavelength. This value of vegetation is very close to those obtained in various studies, such as Humes et al. (1994) and Labed and Stoll (1991). Another surface pattern commonly seen in TM images is man-made building surface, which is mainly composed of concrete materials. Thermal characteristics of man-made building surface are similar to those of bare rocks. Thus we may simply estimate the average emissivity of man-made surface as ε_m=0.970 for convenience of computation.

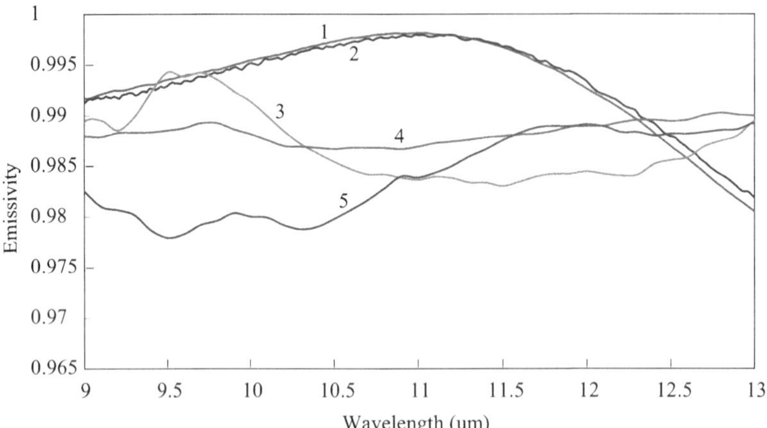

Figure 4.4 Emissivity change of typical waters and vegetation canopies with wavelength, in which curve 1 represents fresh water, 2 sea water, 3 deciduous, 4 conifer, and 5 grass.

4.4 Determination of vegetation fraction

In order to estimate LSE, we have to classify the image pixels into various surface patterns especially water, man-made surface and natural surface. Then we have to estimate the vegetation fraction P_v for both man-made and natural surfaces. Since green index has been widely used for remote sensing of vegetation, we follow this procedure to develop a method here to directly relate the pixels' green index with their vegetation fraction for LSE estimation and LST retrieval. TM data have 7 bands, of which bands 3 and 4 are respectively in red and NIR wavelength. Vegetation has a unique spectral feature of strong absorption in red and high reflection in NIR wavelengths. Therefore, various green indices have been proposed for remote sensing of vegetation. The most widely used green index is NDVI, i.e. normalized difference of vegetation index, computed as follows for Landsat TM data:

$$NDVI=(\mathrm{TM4}-\mathrm{TM3})/(\mathrm{TM4}+\mathrm{TM3}) \tag{4.30}$$

where TM4 and TM3 are DN values of bands 4 and 3 of Landsat TM data. Generally speaking, NDVI value increases with density of vegetation canopy. A fully vegetated pixel will therefore have a maximal NDVI, and a complete bare soil will have minimal NDVI. A pixel with NDVI between these two extremes will have different fractions of vegetation cover and bare soil. Therefore, we can compute a pixel's vegetation fraction as

$$P_v=(NDVI-NDVI_s)/(NDVI_v-NDVI_s) \tag{4.31}$$

in which $NDVI_v$ and $NDVI_s$ are NDVI of fully vegetated pixel and pure bare soil pixel, respectively. When NDVI of a pixel is greater than $NDVI_v$, i.e. $NDVI>NDVI_v$ we have $P_v=1$. Otherwise we have $P_v=0$ if $NDVI<NDVI_s$. It is vital to determine $NDVI_v$ and $NDVI_s$ for vegetation fraction estimation. Generally NDVI of bare soil surface such as desert is very low, usually lower than 0.1. Therefore, we may use this value as $NDVI_s$, i.e. $NDVI_s=0.1$. It has been found that NDVI tends to be saturated at -0.75, which indicates that a very dense vegetation canopy pixel may only have a NDVI value of 0.75. This feature provides us the way to determine the value of $NDVI_v$ as $NDVI_v=0.75$ for vegetation fraction P_v estimation. With this determination, we are able to estimate vegetation fraction in pixel scale for LSE estimation.

We apply the above methodology to Lingxian region, Shandong Province, in North China Plain for LSE estimation and LST retrieval from Landsat TM data (Qin et al., 2006b). Economically the region is characterized with intensive agricultural activities. Scattering with numerous villages, the region has very high density of population distribution, usually up to 300 persons per km^2. Spatial variation of vegetation cover is obvious in the region. The cover is dense in the farmlands and sparse in city and villages.

Using the above methodology, we compute LSE of the region from Landsat TM data acquired on August 21, 2003. Figure 4.5 shows the result of LSE estimation in the region. As seen from Figure 4, the two reservoirs have the highest LSE, which is -0.995. Since the acquisition was carried in the summit of crop growing season, vegetation cover density directly relates to the value of LSE to P_v. Different LSE values can be seen in the farmlands due to their differences in crop canopy cover rate. For those with dense crop canopy, LSE value can be high up to 0.982-0.986. Most pixels have LSE values between 0.979-0.982. The county city Lingxian have very few trees hence the LSE is also low, with only 0.970. Most villages also have similar LSE as the county city.

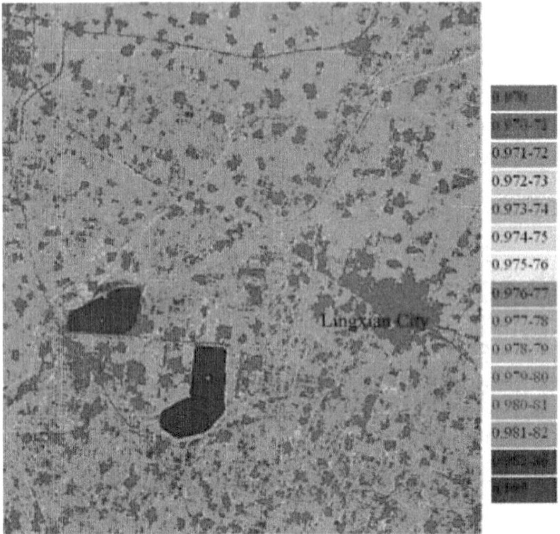

Figure 4.5 Estimated land surface emissivity of Lingxian region in North China Plain, computed from Landsat TM data acquired on August 21, 2003.

Spatial variation of LST in the region is shown in Figure 4.6. We computed the atmospheric parameters with the ground observation of atmospheric factors such as air temperature from meteorological station in the region. The image was acquired at 10:25am when air temperature near the ground was -24°C. Using a typical atmospheric water vapor content of 2.5g/cm^2, we estimated that the effective atmospheric mean temperature Ta in equation (5) was -18.09°C. Thus the atmospheric transmittance was 0.7430 for TM6 wavelength at the time of Lansat satellite over-pass. Since the region is a fluvial plain with a flat feature of topology, we do not consider spatial variation of the two atmospheric parameters for LST retrieval. Figure 5 indicates that effect of heat island is very obvious in the region. High LST can be seen in Lingxian City, where several places have LST of 36-38°C in the hot summer morning. The villages also have high LST, which is 31-35°C, slightly lower than that of Lingxian City. As a contrast, the crop canopy temperature is only in 27-30°C, which is -5°C lower than that of Lingxian City. The lowest LST is seen in the two reservoirs, only with 24-25°C, due to high conductivity of water for heat flux to transfer downward.

Bothe the atmosphere and the ground shapes their impacts on thermal radiance transfer to the remote sensor in space, which makes the remotely sensed surface temperature is far from its true kinetic one. The temperature directly converted from remotely observed thermal radiance is termed as brightness temperature at satellite level. The true surface temperature or the LST can only be retrieved from the remotely sensed thermal radiance using such methodology as mono-window algorithm with consideration of both atmosphere and ground impacts. Figure 3.8 compares the difference of on-board satellite brightness temperature with the retrieved LST. In Lingxian region, the true water surface temperature has the least difference from the observed on-board brightness temperature, usually in 2-3°C, with an average of 2.51°C (Figure 4.7). The difference between LST and brightness temperature is more obvious for the farmland samples, usually in 3-4°C, with an average of 3.52°C. As seen in Figure 6, the greatest difference is observed to be over villages and towns, which is greater than 4.5-7.5°C. Such a big difference between true surface temperature and remotely observed temperature supports the strong necessity to compute LST from thermal remote sensing data.

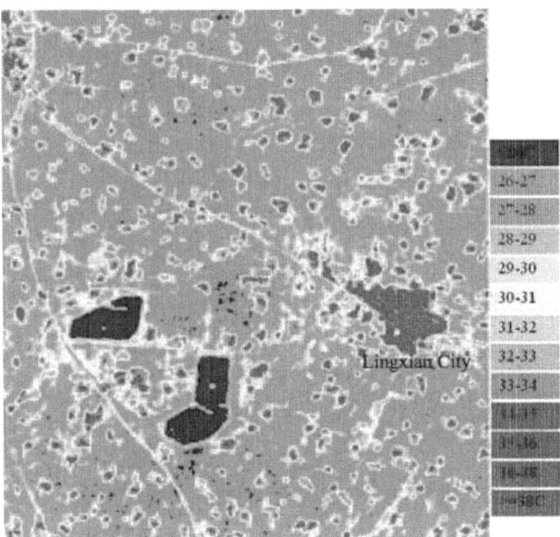

Figure 4.6 Spatial variation of land surface temperature in Lingxian region, retrieved from Landsat TM6 data acquired on August 21, 2003 at about 10:00am.

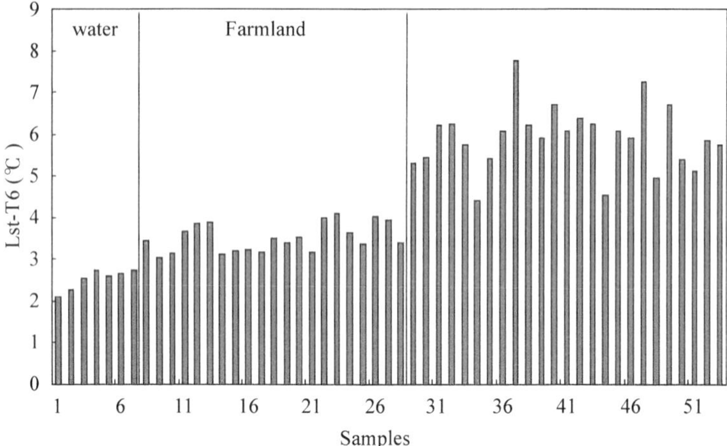

Figure 4.7 Difference between land surface temperature and brightenss temperature over Lingxian region, Shangdong Province, in North China Plain.

4.5 Determination of effective mean atmospheric temperature

Though it is difficult to directly measure the mean atmospheric temperature *in situ* satellite pass, there are several ways to have its estimation for the LST retrieval. Here we intend to follow the method of Sobrino et al. (1991) for the estimation, which relates the determination of T_a with water vapor distribution in the atmospheric profile. According to Sobrino et al. (1991), the effective mean atmospheric temperature T_a can be expressed as:

$$T_a = \frac{1}{w}\int_0^w T_z dw(z,Z) \tag{4.32}$$

where w is total water vapor content in the atmosphere from ground to the sensor altitude Z, T_z is atmospheric temperature at altitude z, $w(z,Z)$ represents water vapor content between z and Z.

Thus, the determination of T_a requires the distribution of atmospheric temperature and water vapor content at each layer of the profile. This is generally unavailable for many studies such as our case. Atmospheric simulation model LOWTRAN 7 provides several standard atmospheric profiles, which can be used to combine with local meteorological data for T_a determination.

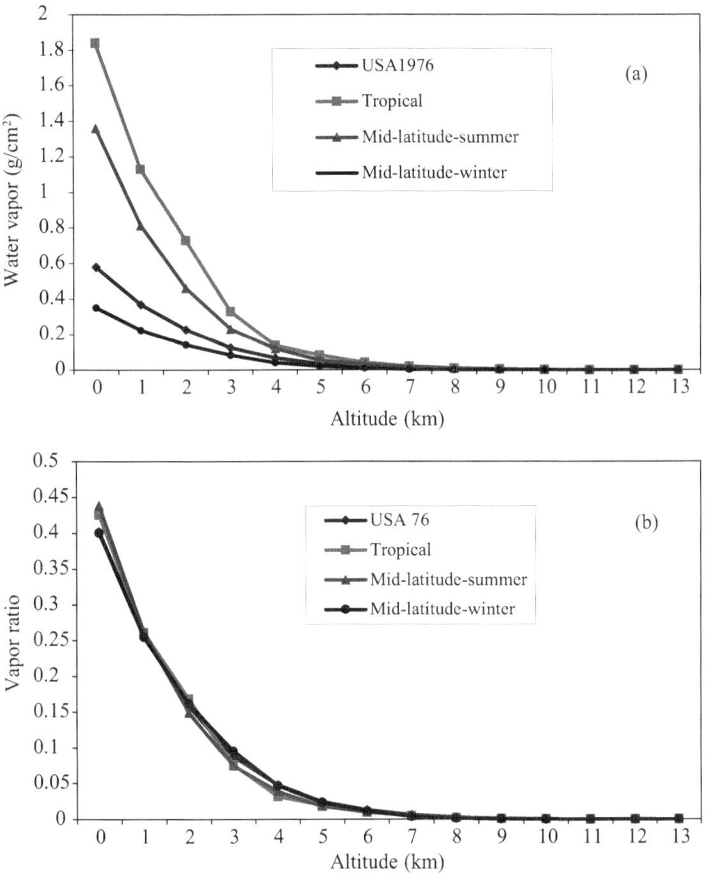

Figure 4.8 Water vapor content (a) and its ratio to the total R_w (b) against the altitude.

In order to determine the effective mean atmospheric T_a, we examine the distribution of water vapor content and atmospheric temperature in the standard atmospheric profiles provided by LOWTRAN 7 model. Figure 4.8 shows the distribution of water vapor content and atmospheric temperature against the altitude. Four general profiles are considered: USA 1976, tropical, mid-latitude summer and mid-latitude winter. It is the fact that most atmospheric water vapor is concentrated in the lower atmosphere (Sobrino et al. 1991) especially in the first 3km of the profile (Figure 4.8a). Though total water vapor content is different (1.44g/cm^2 for USA 1976, 4.33g/cm^2 for tropical), the distribution of water vapor in the atmospheric profiles is very similar (Figure 4.8b). The first layer contains about 40.206% of the total water vapor content in USA

1976 profile, about 43.35% in mid-latitude summer profile (Figure 4.8b). Details of the distribution are given in Table 3.2 for the profiles. Therefore, using an average of the four profiles, we can develop a standard distribution for estimation of water vapor content at each layer of the profile as follows:

$$w(z)=wR_w(z) \tag{4.33}$$

where $w(z)$ is water vapor content at altitude z, $R_w(z)$ is the distribution of water vapor in the atmospheric profile, given in the average column of Table 4.3. As indicate by Figure 4.8a and Table 4.3, water content at the upper layers (>10km) is negligibly small. And at the sensor altitude, we can rationally assume $w(Z)=0$.

The finite term of equation (4.30) at altitude z can be expressed as water vapor content of the layer, *i.e.* $dw(z,Z)=w(z)$. Thus, equation (4.30) can be rewritten as

$$T_a=\frac{1}{w}\sum_{z=0}^{m}T_z w(z) \tag{4.34}$$

where m is number of the atmospheric layers under consideration. Because $w(z)$ in the upper atmospheric layers is very small, the effective mean atmospheric temperature is mainly determined by T_z in the lower atmospheric layers.

Table 4.3 Radio of water vapor content to the total in different atmospheric profiles

Altitude (km)	USA 1976	Tropical	Mid-latitude summer	Mid-latitude winter	Average $R_w(z)$
0	0.402058	0.425043	0.438446	0.400124	0.416418
1	0.256234	0.261032	0.262100	0.254210	0.258394
2	0.158323	0.168400	0.148943	0.161873	0.159385
3	0.087495	0.075999	0.074471	0.095528	0.083373
4	0.047497	0.031878	0.038364	0.046510	0.041062
5	0.024512	0.019381	0.017925	0.023711	0.021382
6	0.012846	0.009771	0.009736	0.011514	0.010967
7	0.006250	0.004782	0.005223	0.004092	0.005087
8	0.003132	0.002257	0.002611	0.001471	0.002368
9	0.001049	0.000954	0.001315	0.000587	0.000976
10	0.000358	0.000349	0.000616	0.000238	0.000390
11	0.000142	0.000104	0.000185	0.000060	0.000123
12	0.000055	0.000032	0.000044	0.000026	0.000039
13	0.000023	0.000008	0.000009	0.000016	0.000014
14	0.000009	0.000004	0.000004	0.000011	0.000007
15	0.000006	0.000002	0.000002	0.000008	0.000004

It is well known that atmospheric temperature decreases with altitude in lower atmosphere. Figure 4.9 perfectly depicts the change of atmospheric temperature with altitude. Even though the atmospheric temperature of the four profiles differs greatly at the ground surface, it seems that the attenuation of atmosphere makes it very closed to each other at about 13km height. Atmospheric temperature at this altitude changes from 216.7K in USA 1976 through 217K in tropical to 218.2K in mid-latitude winter profile with an average of 217K. With this property, we can compute the attenuation rate of atmospheric temperature for the four profiles as Table 3.3, from which a similar attenuation rate can be seen in the four profiles. Thus, T_z for each layer can be estimated as follows:

$$T_z=T_0-R_t(z)(T_0-217) \tag{4.35}$$

where T_0 is atmospheric temperature at the ground, $R_t(z)$ is the attenuation rate of temperature at z, given in the average column of Table 4.4. Accordingly, provided that T_0 is given, the atmospheric temperature distribution at each layer can be directly computed from equation (4.35).

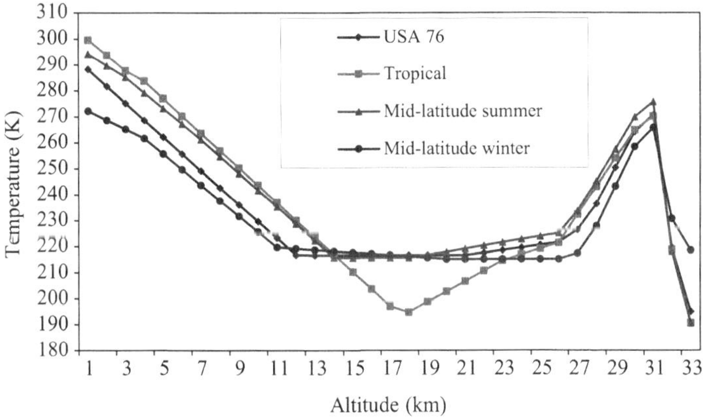

Figure 4.9 Distribution of atmospheric temperature against the altitude

Table 4.4 Attenuation rate of atmospheric temperature in different profiles

Altitude (km)	USA76	Tropical	Mid-latitude summer	Mid-latitude winter	Average $R_t(z)$
0	0	0	0	0	0
1	0.0912921	0.0725514	0.0582902	0.0634058	0.0713849

（续表）

Altitude (km)	USA76	Tropical	Mid-latitude summer	Mid-latitude winter	Average $R_t(z)$
2	0.1825843	0.1451028	0.1165803	0.1268116	0.1427697
3	0.2738764	0.1934704	0.1943005	0.1902174	0.2129662
4	0.3651685	0.2744861	0.2720207	0.298913	0.3026471
5	0.4564607	0.3555018	0.3497409	0.4076087	0.392328
6	0.5477528	0.4365175	0.4274611	0.5163043	0.482009
7	0.6390449	0.5163241	0.511658	0.625	0.5730068
8	0.7303371	0.5973398	0.5958549	0.7336957	0.6643069
9	0.8216292	0.6783555	0.6800518	0.8423913	0.755607
10	0.9115169	0.758162	0.7629534	0.951087	0.8459298
11	1.002809	0.8415961	0.8471503	0.9601449	0.9129251
12	1.0042135	0.9201935	0.9313472	0.9692029	0.9562393
13	1.0042135	1	1.015544	0.9782609	0.9995046
14	1.0042135	1.0810157	1.0168394	0.9873188	1.0223469
15	1.0042135	1.1608222	1.0168394	0.9963768	1.044563

Therefore, when the local atmospheric profile of water vapor and atmospheric temperature is not available *in situ* satellite pass, the following procedure can be used to determine T_a for LST retrieval from Landsat TM6 data.

(1) Use available atmospheric profiles of the study region to compute $R_w(z)$ and $R_t(z)$ for atmospheric temperature and water vapor content at each layer and use the average of the rates to represent the standard atmospheric distribution of the region. If the profiles are also unavailable (such as our case), the average columns of Table 4.3 and 4.4 can serve as the standard distribution.

(2) Use equation (4.31) and the total atmospheric water vapor content to fit into the standard distribution of atmospheric water vapor content for computation of water vapor content at each layer of the profile. If total atmospheric water vapor content is not available, the water vapor content at the near surface $w(0)$ can be used to estimate the total water content as $w=w(0)/R_w(0)$.

(3) Use equation (4.33) and the known local air temperature at the near surface (usually 2m above ground) as T_0 to fit into the standard atmospheric temperature

distribution for computation of atmospheric temperature at each layer of the profile.

(4) Use equation (4.32) to compute the effective mean atmospheric temperature T_a through summation of atmospheric temperature multiplying water vapor content for each layer of the profile.

Once T_a has been calculated, the LST can be retrieved using the mono-window algorithm described in equation (4.32).

4.6 Determination of atmospheric transmittance

Atmospheric transmittance is a critical parameter required for LST retrieval from Landsat TM data. Generally speaking, the determination of atmospheric transmittance for channel 6 of Landsat TM is undertaken through the simulation of atmospheric conditions with LOWTRAN or MODTRAN program. Due to its extremely importance in influencing the variation of atmospheric transmittance, water vapor content has been extensively used as the determinant of estimating atmospheric transmittance (Sobrino et al., 1991; Coll et al., 1994a; Cracknell, 1997).

In the study, LOWTRAN 7 program is used to simulate the relation of water vapour content to atmospheric transmittance. The general water vapour range of 0.4-4.0g/cm^2 is considered for the simulation under two profiles: low and high temperature profiles. The atmospheric temperature near the surface for the high temperature profile is defined as 35°C and for the low one, 18°C. The swath of Landsat TM is about 185km, which results in a viewing zenith angle of about 6° for its edge pixels. Thus, we consider an average zenith angle of 3° for the simulation of atmospheric impact on transmittance. Simulation results are shown in Figure 4.10, from which we can see that TM6 transmittance decreases steadily with water vapour content increase. Atmospheric transmittance of TM6 is high up to above 0.9 for water vapour content less than 0.8g/cm^2. The transmittance decreases to about 0.8 at 2.0g/cm^2 and 0.7 at 2.5g/cm^2. It may be lower than 0.5 for water vapour content greater than 4.0g/cm^2. Another feature shown in Figure 4 is the transmittance difference between the two profiles. The difference is very small when water vapour content is low. However, it increases rapidly with the water vapour content. High temperature profile has higher transmittance than low temperature profile for the same water vapour content. The transmittance difference between high and low temperature profiles is about 0.007 at water vapour content 1.0g/

cm^2. The difference increases to 0.031 at 2.0g/cm^2, 0.0558 at 3.0g/cm^2 and 0.0799 at 4.0g/cm^2. The change of transmittance with water vapour is not linear for the whole range 0.4-4.0g/cm^2 but for a small segment the relationship is close to linearity. This characteristic provides the possibility of establelishing some simple linear equations to estimate transmittance from water content for Landsat TM6 operating in 10.5-12.5μm. An effort of establelishing the estimation equations is given in Table 4.5 for the range 0.4-3.0g/cm^2, which is the general case in the world.

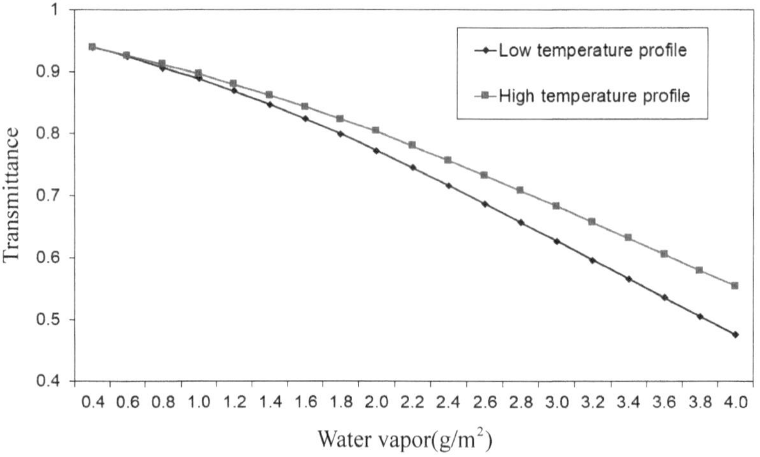

Figure 4.10 Change of atmospheric transmittance with water vapour content in mid-latitude region.

Table 4.5 Estimation of atmospheric transmittance for Landsat TM channel 6

Profiles	Water vapour (w) (g/cm^2)	Transmittance estimation equation	Squared correlation R^2	Standard error
High air temperature	0.4-1.6	τ_6=0.974290−0.08007w	0.99611	0.002368
	1.6-3.0	τ_6=1.031412−0.11536w	0.99827	0.002539
Low air temperature	0.4-1.6	τ_6=0.982007−0.09611w	0.99463	0.003340
	1.6-3.0	τ_6=1.053710−0.14142w	0.99899	0.002375

The estimation equations listed in Table 4.5 have high squared correlation and low standard error, which means that the estimation of transmittance with water vapour content by these equations will be in a high accuracy. For a possible measurement error of 0.2g/cm^2 (this is the general accuracy of water vapour content estimation from sunphotometer measurements), the possible maximal estimation error of atmospheric transmittance for Landsat TM6 is <0.029. An accurate estimation of transmittance, as indicated in following section, is very important in retrieval of LST from Landsat TM6 data.

4.7 Sensitivity analysis of the mono-window algorithm

Compared with split window algorithm for AVHRR, the mono-window algorithm for Landsat TM requires three critical variables to estimate LST: ground emissivity, atmospheric transmittance and effective mean atmospheric temperature. Due to many difficulties such as unavailability of precise profile data, the determination of these three variables will unavoidably involve some errors. In order to analyze the impact of the possible estimation error of these critical variables on the LST retrieval, sensitivity analysis is necessary. For convenience, equation (3.33) is used to express the possible LST estimation error.

Sensitivity analysis is performed under several conditions. First of all, natural surfaces generally have an emissivity of about 0.95-0.98 in the thermal wavelength 10-13μm (Givri, 1995; Coll and Caselles, 1994; Humes et al., 1989; Becker, 1987). And the emissivity of 0.97 is used for the sensitivity analysis. Second, the clear atmospheric profile in the arid environment usually has an average water vapor content of about 1.5-2.0g/cm^2, which corresponds to transmittance of about 0.77-0.85. Thus, the transmittance of 0.80 is used for the analysis. The time for Landsat 5 to pass the region was at about 9:30-10:00am local time when the atmospheric temperature is generally not very high. Thus, the effective mean temperature of 15°C is used. The air temperature near the surface corresponding to this mean temperature is about 25°C for water vapor content of about 2.0g/cm^2. Under these conditions, the sensitivity analysis of the algorithm is performed for ground emissivity error, transmittance error and effective mean atmospheric temperature error.

4.7.1 Effect of ground emissivity

Figure 4.11 illustrates the probable LST estimation error due to the possible ground emissivity error. Several important features can be seen in Figure 3.16a, which shows the LST estimation error against brightness temperature of TM6 for the four possible emissivity errors. δT decreases with brightness temperature when it is below 13°C, which is an equilibrium point. From this point, δT increases rapidly with

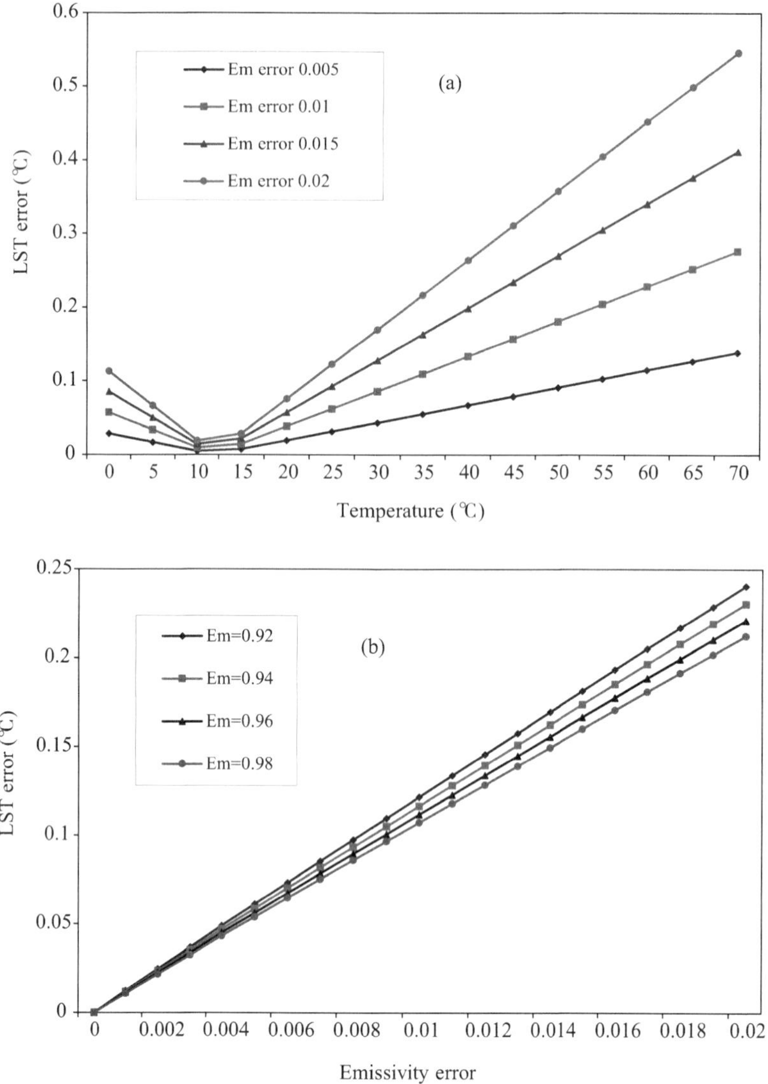

Figure 4.11 Probable LST estimation error due to the possible emissivity error with (a) LST error against brightness temperature and (b) average LST error against emissivity error.

temperature for all cases. Linear correction between LST error and temperature level is also very clear in all cases. According to these relationships, for an emissivity error of about 0.01, we can expect δT less than 0.3°C in the temperature range 0-70°C and for the emissivity error of 0.02, the LST error is less than 0.6°C in the same temperature range (Figure3.16a). This LST estimation error corresponding to emissivity error is quite small in comparison with the split window algorithm for AVHRR. Generally, the estimation of ground emissivity in the range of 10-13μm can reaches the accuracy of less than 0.02. Provided this assumption, the mono-window algorithm can produce quite accuracy in LST estimation from Landsat TM data.

Since most surfaces have temperature above 10°C, an average δT in the temperature range of 10-55°C has been plotted against the possible ground emissivity error (Figure 3.16b). Four ground emissivity cases are plotted in Figure 3.16b, which indicates that the LST error has very small change with the ground emissivity. This means that the probable LST error is only sensible to emissivity error but not sensible to the level of ground emissivity itself. For the maximal emissivity error of 0.02, the LST error difference is only about 0.02°C between τ_6=0.92 and τ_6=0.98. Though the δT increases steadily with emissivity error, the average δT is less than 0.25°C for the maximal emissivity error 0.02 (Figure 3.16b). Therefore, it can be concluded that the algorithm can provide quite accurate LST estimation in spite of some emissivity error.

4.7.2 Effect of atmospheric transmittance

In contrast with the insensible response to emissivity error, the algorithm is quite sensible with transmittance error. Figure 4.12 shows the probable LST estimation error due to possible transmittance error. Again, an equilibrium point exists at brightness temperature level of about 13°C (Figure 4.12a) where δT is 0 for various transmittance errors. When brightness temperature level is below this equilibrium point, LST estimation error is less than 0.25°C for $\delta\tau_6 \leqslant 0.02$. However, δT increases rapidly with temperature and transmittance. For $\delta\tau_6$=0.02, δT is about 0.577°C at temperature level T=30°C and it increases to about 1.058°C at the level T=45°C and 1.378°C at the level T=55°C (Figure 4.17a). For the maximal $\delta\tau_6$=0.05, δT may reach above 3°C at the temperature level T>45°C. Therefore, an accurate estimation of atmospheric transmittance is much more important in using the algorithm for LST retrieval than the emissivity estimation.

Moreover, δT is also very sensible to the change of atmospheric transmittance (Figure 4.12b). Average LST error in the temperature range of 10-55°C is plotted against transmittance error for several transmittance levels. When transmittance is high, the increase of average δT with $\delta\tau_6$ is much slower. Specifically, average δT is about 0.74-0.96°C for $\delta\tau_6$=0.02 when τ_6 is 0.7-0.8. For the same $\delta\tau_6$, the average δT increases to 1.29-1.84°C when τ_6 decreases to 0.5-0.6. Generally speaking, the accuracy of atmospheric water vapor measurement is within 0.2g/cm^2. Consequently it produces an error of 0.016-0.28 in τ_6 estimation. Therefore, we can expect that the algorithm can provide an accuracy of LST estimation within 1°C for transmittance higher than 0.7. This accuracy is generally accep Table for most study purposes using Landsat TM data.

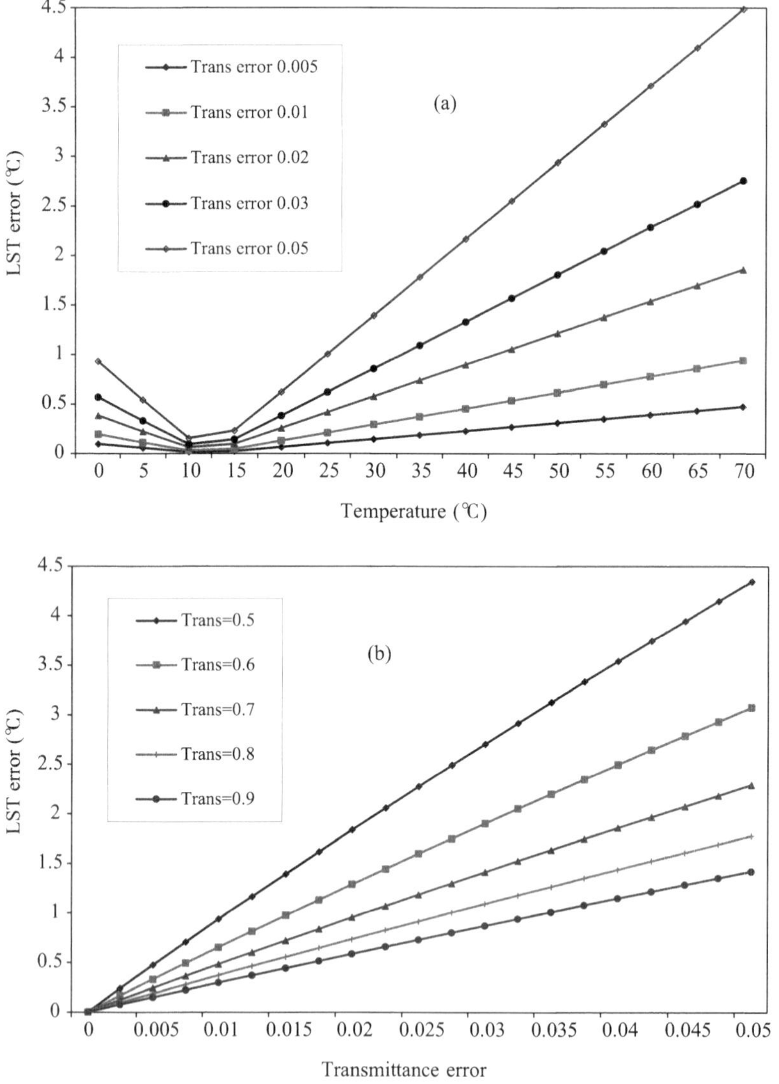

Figure 4.12 Probable LST estimation error due to atmospheric transmittance error with (a) LST error against brightness temperature and (b) average LST error against transmittance error.

4.7.3 Effect of mean atmospheric temperature

Sensitivity analysis of the algorithm to effective mean atmospheric temperature Ta indicates that the δT does not change with brightness temperature and the mean atmospheric temperature itself but does change linearly with the possible error of the mean air temperature. This can be understood through the derivation of the LST estimation error according to equation (3.33). For a small error of T_a, actually we can derive that

$$\begin{aligned}\delta T &= |T_s(T_a+\delta T)-T_s(T_a)| \\ &= |\{[a_6(1-C_6-D_6)+(b_6(1-C_6-D_6)+C_6+D_6)T_6-D_6(T_a+\delta T_a)]/C_6\}- \\ &\quad \{[a_6(1-C_6-D_6)+(b_6(1-C_6-D_6)+C_6+D_6)T_6-D_6T_a]/C_6\}| \\ &= |[D_6T_a-D_6(T_a+\delta T_a)]/C_6| \\ &= |(D_6/C_6)\delta T_a|\end{aligned}$$

For a given emissivity and transmittance, D_6 and C_6 are fixed as constant according to equation (4.15) and (4.16). Therefore, δT only varies with δT_a but not T_6 and T_a itself. In order to analyze the change of δT with δT_a, I compute the ratio of D_6 to C_6 for several combinations. The results are given in Table 4.6, which indicates that the ratio D_6/C_6 is mainly depended on transmittance. Emissivity has slightly effect on the ratio. For the transmittance of 0.8, the ratio changes from 0.259 for emissivity of 0.98 to 0.279 for emissivity of 0.94 with an average of 0.269. For the transmittance of 0.7, the ratio changes in the range of 0.443-0.475 for the emissivity range of 0.94-0.98. Based on the average ratio of the combinations for different transmittances, the change of δT against δT_a is plotted in Figure 4.13.

Table 4.6 Comparison of the ratio D6/C6 for different combinations

Emisivity	Transmittance	Parameter C_6	Parameter D_6	Ratio D_6/C_6	Average ratio
0.94	0.6	0.564	0.4144	0.734752	
0.95	0.6	0.57	0.412	0.722807	
0.96	0.6	0.576	0.4096	0.711111	0.711352
0.97	0.6	0.582	0.4072	0.699656	
0.98	0.6	0.588	0.4048	0.688435	
0.94	0.7	0.658	0.3126	0.475076	
0.95	0.7	0.665	0.3105	0.466917	
0.96	0.7	0.672	0.3084	0.458929	0.459093
0.97	0.7	0.679	0.3063	0.451105	
0.98	0.7	0.686	0.3042	0.44344	

（续表）

Emisivity	Transmittance	Parameter C_6	Parameter D_6	Ratio D_6/C_6	Average ratio
0.94	0.8	0.752	0.2096	0.278723	
0.95	0.8	0.76	0.208	0.273684	
0.96	0.8	0.768	0.2064	0.26875	0.268852
0.97	0.8	0.776	0.2048	0.263918	
0.98	0.8	0.784	0.2032	0.259184	
0.94	0.9	0.846	0.1054	0.124586	
0.95	0.9	0.855	0.1045	0.122222	
0.96	0.9	0.864	0.1036	0.119907	0.119955
0.97	0.9	0.873	0.1027	0.11764	
0.98	0.9	0.882	0.1018	0.11542	

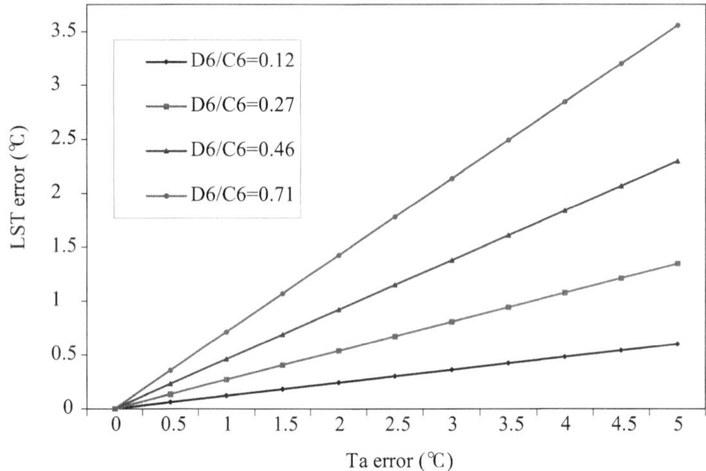

Figure 4.13 Probable LST estimation error due to the possible mean atmospheric temperature error.

Figure 4.13 indicates that LST estimation error increases rapidly with the possible mean atmospheric temperature error and the average ratio D_6/C_6, which is mainly depended on transmittance (Table 4.6). When mean atmospheric estimation error is about 1°C, the probable LST estimation error is about 0.27°C for the average ratio D_6/C_6=0.27 which is corresponding to τ_6=0.8. However, the δT may reach the level of high up to 0.71°C for the same δT_a but the ratio D_6/C_6=0.71 or τ_6=0.6. If δT_a is greater than 2°C, the δT is about 0.54°C for τ_6=0.8 and about 0.91°C for τ_6=0.7.

Therefore, it can be concluded that an accurate estimation of mean atmospheric temperature is extremely important for reaching an accurate LST retrieval from only one thermal channel of Landsat TM data. An obvious LST estimation error (δT>1°C) is generally unavoidable for τ_6<0.65 and δT_a>2°C. However, if the sky is very clear so that τ_6>0.8, the possible δT will be less than 1°C for δT_a high up to 4°C. And such a big δT_a rarely happens for most cases. Therefore, a clear sky with lower water vapor content is an ideal atmospheric condition for remote sensing of LST with Landsat TM data. In our arid region, the water vapor is generally less than 2.0g/cm^2, which means that τ_6>0.78. With this transmittance, the possible LST error is expected to be less than 0.53°C for δT_a=2°C and less than 0.81°C for δT_a=3°C.

4.8 Spatial distribution of LST in the region

The above mono-window algorithm was used to retrieve LST from available Landsat TM6 data of the region. Figure 4.14 showed the result of such a retrieval effort, i.e. the spatial variation of LST distribution in the border region. This image was taken in March 29, 1995 when it is the bloom season of the year. The satellite Landsat-5 passed the area at about 9:30 am. A clear difference of LST on both sides can be seen in this image. Generally speaking, the Israeli side has obvious higher LST. On average, LST is 33.87°C on the Israeli side and 31.01°C on Egyptian side. Thus the difference is about 2.86°C. High LST can be seen on the Israeli side in the right upper corner of the image. The LST in this area is high up to 36-38°C. Low LST concentrates in the middle part of the Egyptian side. The LST in this part is only about 28-29°C. Thus, maximal difference of LST is high up to 8-9°C on both sides. In the area close to the border, LST is at the middle level, vibrating at about 31-32°C on the Israeli side and 29-30°C on the Egyptian side. A sharp contrast of LST still can be seen in this area adjacent to the border on both sides even though the difference is lower than the average one (Figure 4.14).

Considered more vegetation cover on the Israeli side and the booming season, the higher LST on the Israeli side is really an interesting anomaly. The only reason is that the Israeli side has much higher biogenic crust cover while the Egyptian side has much more sand surface. Biogenic crust cover on the Israeli side is estimated to be high up to about 72% and sand surface on the Egyptian side is above 80%. Due to low albedo, the biogenic crust usually has much higher surface temperature than the sand (see next chapter). Ground truth measurements indicate that surface temperature on biogenic crust is above 3°C higher than on sand in summer. Therefore, the higher LST on the Israeli side is due to the contribution of biogenic crust overcoming the cooling process of its more vegetation cover. Examination of this mechanism will be given in chapter 9.

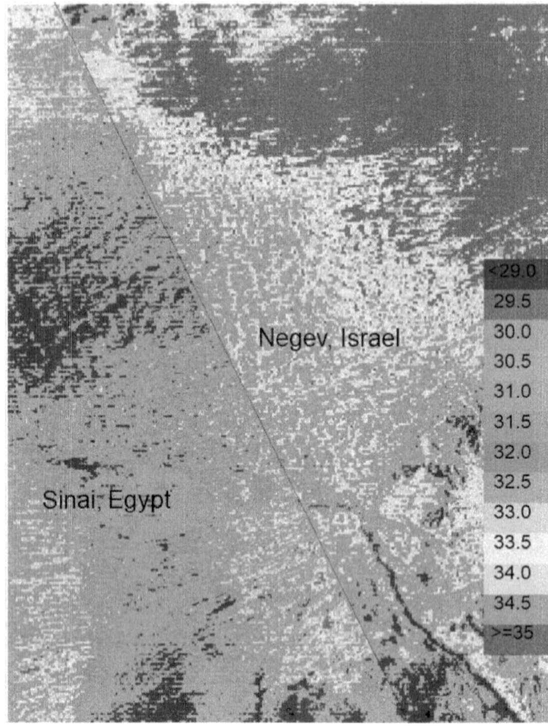

Figure 4.14 Land surface temperature variation in the sand dunes across the Israel-Egypt border, retrieved from Landsat TM6 data of 29 March, 1995

Another LST image retrieving from Landsat TM data is Figure 4.15. This image was taken at about 9:45 on September 21, 1995. The spatial variation of LST distribution revealed in this image is similar with that shown in Figure 3.19. Obvious LST difference can be clearly seen on both sides. The Israeli side has LST of 37-39°C while the Egyptian side only has 35-37°C. The average LST is 38°C on the Israeli side and 36°C on the Egyptian side. Therefore, the difference is about 2°C. High LST can be seen on the right top of the image, where LST is up to above 38°C. Low LST can be found on the bottom part of the image, where the landscape is mainly rocky terrain patterns.

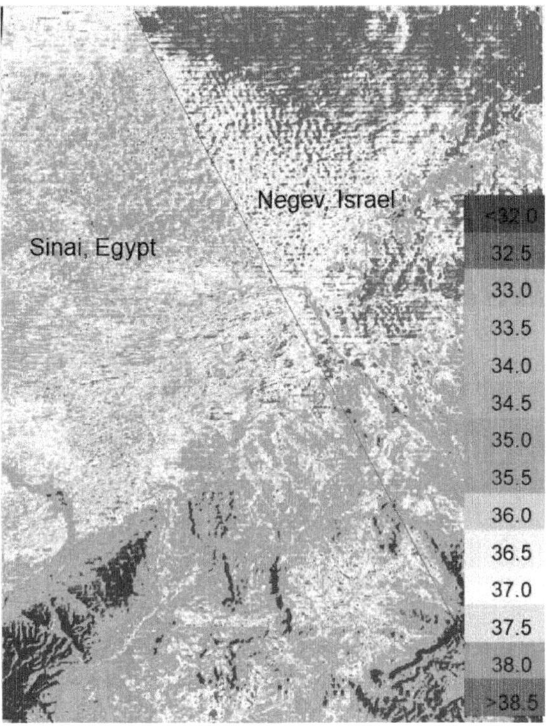

Figure 4.15 LST variation in the sand dunes across the Israel-Egypt border, retrieved from Landsat TM data of September 21, 1995.

Several other TM images about this region are also available for the analysis. Though these images do not have a full cover of the whole region, they still can be used to analyze the LST change of the region. After LST retrieval, the average LST on both sides was calculated for each image and the result is shown in Figure 4.16, which depicts the average LST change and its difference on both sides. Several features can be seen in this graph. In hot dry season, the LST of the region was high up to about 50°C in morning at about 10:00 am. In early and late dry seasons the LST is about 40°C. However, it is very low (about 13-20°C) in cool wet season. LST difference is very obvious on both sides in dry and late wet seasons. Usually the Israeli side has an average LST of above 2°C higher than the Egyptian side. The sharp LST difference disappears in cool wet season when temperature is below 20°C. The anomalous thermal variation on both sides is governed by the seasonal change of local climatology, which couples with the moisture vibration in the soil and the air. When temperature is lower than some level such as 20°C, the Israeli side will not have higher LST than the Egyptian side. Rainfall has very important impact on performance of the anomalous LST change in the region. There was a very heavy raining process in November 2-6, 1994 and the total rainfall was high up to 54mm. Several small rains also occurred between November 10 and 18, 1994, with total rainfall of 7.1mm. Therefore, when Landsat passed the region on November 21, 1994, the ground surface was very wet. In this case, LST of the region was below 20°C and the Egyptian side had relatively hotter though it is very weak (Figure 4.16)

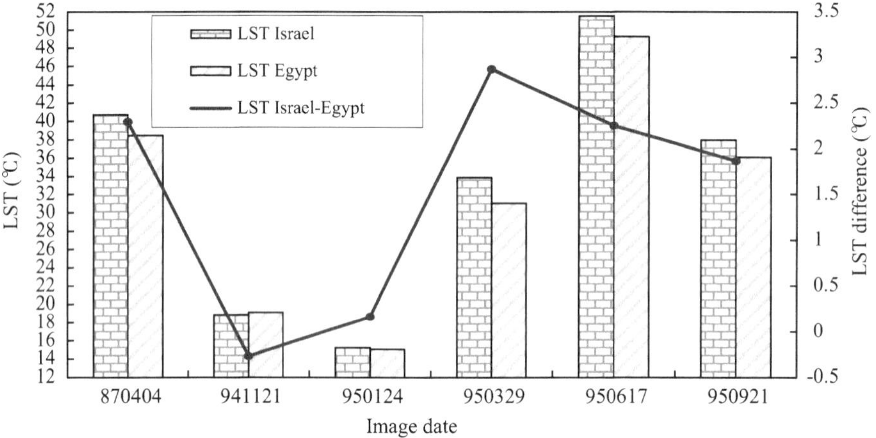

Figure 4.16 Average LST change on both sides and their difference, retrieving from Landsat TM images.

5 Ground truth measurements of ground emissivity and surface temperature

Based on the LST retrieval from NOAA-AVHRR and Landsat TM data, Temporal and spatial variation of LST in the border region has been analyzed in the previous chapter. This remote sensing analysis indicates that the border region does have a sharp contrast of LST change during the daytime, even though the phenomenon behaves fluctuation in different seasons. In order to validate the observed LST contrast on remote sensing data, we need to determine the ground emissivity of the typical surface patterns and perform the ground truth measurements of surface temperature change in these patterns. Based on these measurements, mixed surface temperature change on both sides can be analyzed after considering their surface composition difference. The measurements can also provide some information about the possible level of surface temperature change in the region for further micro-meteorological modeling and analysis.

5.1 Experiments for ground emissivity measurement

Emissivity is an important factor that affects the observed surface temperature change through detecting the thermal radiance of the object under study. The ground surface is not a black body that has a full capability to emit its radiant energy under specific temperature. All natural objects are able to emit only part of its potential radiant energy (Caselles et al., 1997). Therefore, the direct output of temperature measurement by an IR thermometer or a remote sensing system can only give the radiant temperature change of the objects. In order to know the truth or kinetic temperature change of the ground, surface emissivity has to be considered for the correction of the observed radiant temperature. Moreover, the LST retrieval of remote sensing data in the previous chapter also requires the known of surface emissivity. These requirements make the determination of ground emissivity necessary in the study.

Though several methods have been proposed (Humes et al., 1994; Labed & Stoll,

1991 a), directly measuring emissivity of the ground surface in the field is still very difficult due to the cost and the instrumentation system. The general way for emissivity determination in an accep Table economic level is to take the soil samples from the field and carry out experiments for the determination. And we also follow this in the study. The soil samples for the experiments were taken from the field during the wet season of 1997/98 using metal cans. We took the sampling for several times: on January 16 and 26, and March 19 and 26, 1998. Great cares were exerted when taking the samples in order to avoid soil structure change. The samples were air-dried on the shelf of our laboratory to compare the surface condition in the dry season of the region.

Biogenic crust and bare sand are the most important surface compositions of the region and together they account for above 85% of the area. The emissivity of biogenic crust on the surface of sandy soil has not been yet reported. Though several studies such as Sobriono and Caselles (1991) and Sabins (1996) stated that sand (mainly quartz) has an emissivity of 0.91-0.92, the emissivity of fine sand in the region, which is mainly composed of dust and rocky particles (Gerson et al., 1985), has also not been studied. Therefore, we focus our experiments on determining the emissivity of these two surface patterns. The emissivity of vegetation has been reported (Humes et al., 1994; Labed & Stoll, 1991 a) and that of Playa, due to its composition, can be viewed as having similar soil properties with silt and clay.

The principle of determining emissivity through laboratory experiments is to compare the difference of radiant temperature between the measured samples and their reference black body. The experiments were done through the use of a water bath with constant temperature inside and an IR thermometer operating in 8-14μm to measure the radiant temperature of the samples. A circulator was installed in the water bath to keep the temperature constant. On the top of the bath is a moveable semitransparent cover so that the samples in the bath can be seen. The cover also acts as an isolated layer to prevent the impact of thermal radiation from the environment on the samples. There is a small hole in the cover to mount the sensor of the thermometer for measuring the radiant temperature of the samples under study.

The soil samples were arranged to float on the water surface of the bath. Five typical sand samples and five typical biogenic crust samples were selected for the measurement. At the same time, a reference with black surface was also put in the bath. Then a semi-transparent cover was used to cover the bath to isolate the bath from environment. The IR thermometer was mounted on the small hole of the cover. A datalogger with storage was connected to the thermometer for data recording. After the samples were put in the bath for more than 2 hours, we started to measure the temperatures of the samples and the reference through operating the cover with the thermometer over them. The output of the thermometer was recorded by a datalogger. Data output frequency of the thermometer was set into one reading per second. The thermometer has ability to react with the thermal radiation in 2-3 seconds. For each sample, we took the measurement for 10-20 seconds so that we would have about 15 temperature readings of the sample. The samples were measured one by one and we repeated the measurements for five times. Three temperature treatments (25 °C, 35 °C and 45 °C) were used for the experiments in order to represent three circumstances of surface temperature change in the region. In summer, the surface temperature of the region is usually high up to above 45 °C and in winter it is about 25 °C for most days.

The emissivity of the samples can be determined on the basis of the assumption that the samples and the black body reference have the same kinetic temperature. Because the thermometer was mounted just on top of the bath with a distance of less than 15 cm from the samples and a cover was used to isolate the bath from its environment, the possible effects of environmental emittance from air and the laboratory walls in detecting their surface temperature can be minimized hence ignored. In order to get an average measured the temperatures of the samples and the black reference were measured in an iterative way for five times. Consequently, the observed difference of the radiant temperatures among the samples can be assumed to be the direct result of their emissivity difference.

5.2 Computation of ground emissivity

The IR thermometer measures the radiant temperature of the samples by detecting their thermal radiance. Actually the thermal radiance reaching the IR sensor is not only the emittance from the sample, but also the effects of the environment. Because the samples are not black bodies with an emissivity of 1, they have ability to reflect the downward thermal radiance from the air and the indoor laboratory wall. And the air between the sensor and the sample surface also has ability to emit some thermal radiance to reach the IR sensor. Thus we have total thermal radiance reaching the IR sensor as

$$I=I_s+I_r+I_a^{\uparrow} \tag{5.1}$$

where I is the total thermal radiance reaching the sensor; I_s is the thermal emittance from the sample; I_r is the reflected downward thermal radiance from air, and laboratory walls and ceiling; and $I_a^{\uparrow}$ is the upward air emittance. Since we used a cover to isolate the bath from the environment, we successfully minimized the effects of thermal radiance from laboratory walls and ceiling on the measurements. Now we only need to consider the downward and upward air emittances. Because our sensor is only about 15cm from the sample surface, the air transmittance τλ within the thin layer can be termed as τλ≈1. Accordingly, both upward and downward air emittances can be computed as

$$I_r=(1-\varepsilon)I_a^{\downarrow}=(1-\varepsilon)(1-\tau_\lambda)\, B_\lambda(T_a^{\downarrow})\approx 0 \tag{5.2}$$

$$I_a^{\uparrow}=(1-\tau_\lambda)B_\lambda(T_a^{\uparrow})\approx 0 \tag{5.3}$$

where $B_\lambda(T_a)$ is Planck radiance function with wavelength λ and temperature T_a. Thus, we have $I\approx I_s$ in our case.

Total emittance of a black body with temperature T_b can be estimated as Stefan-Boltzmann function or Planck function. With Stefan-Boltzmann function, the emittance can be computed as

$$I(T_b)=\sigma T_b^{\,4} \tag{5.4}$$

where $I(T_b)$ is the emittance of the body at temperature T_b, σ is Stefan-Boltzmann constant with σ=5.67×10-8Wm-2K-4. With Planck function, we have

$$I(T_b)=\frac{1}{\lambda_2-\lambda_1}\int_{\lambda_1}^{\lambda_2}\frac{c_1}{\lambda^5(e^{c_2/(\lambda T_b)}-1)}f(\lambda)d\lambda \tag{5.5}$$

where c_1 and c_2 are the first and the second radiance constants with $c_1=hc^2=5.95522012\times10^{-17}$ W m^2 and $c_2=h\,c/k=1.43876869\times10^{-2}$ m K, λ is the wavelength, and $f(\lambda)$ is the spectral response function of the instrument within its wavelength range between λ_1 and λ_2. Since Stefan-Boltzmann function is about the entire wavelength and our instrument is only for the wavelength of 8-14μm, we prefer to select Planck function for the determination of emissivity. The soil sample with temperature To is not black body and only has an emittance of $I(T_o)=\varepsilon I(T_b)$. Thus, its emissivity can be computed as

$$\varepsilon = \frac{I(T_o)}{I(T_b)} = \frac{\int_{\lambda_1}^{\lambda_2} \frac{c_1}{\lambda^5 (e^{c_2/(\lambda T_o)} - 1)} f(\lambda) d\lambda}{\int_{\lambda_1}^{\lambda_2} \frac{c_1}{\lambda^5 (e^{c_2/(\lambda T_b)} - 1)} f(\lambda) d\lambda} \tag{5.6}$$

where T_b and T_o are the measured temperature (K) of black body and the sample.

Black body is only a theoretical concept and there is no absolute black body in nature. Thus, we assume that the black body reference has an emissivity of 0.999. Used this assumption, the temperature of the black body reference is firstly calibrated into its real kinetic temperature. Another consideration in the determination is the possible error resulted from both instruments and operation. Because the bath is set for constant temperature, the true temperature of the black body reference should be the same as the constant temperature in the bath. Usually there is a slight difference from the expected constant temperature. Therefore, calibration has to be done before the above formula is used to calculate the emissivity. In practice, the difference between the temperature of the black body reference and the expected constant temperature is used as the calibration coefficient.

5.3 Results and analysis of ground emissivity measurements

The results of emissivity determination are shown in Figure 5.1 and the detailed calculation is given in Table 5.1 for the treatments of 25 °C and 45 °C. Figure 5.1 indicates that biogenic crust has higher emissivity than sand. According to the experiments, the average emissivity of the five samples at 25 °C is 0.9671 for biogenic crust and 0.9435 for sand. The difference is about 0.0236. The average emissivity at 35 °C is 0.9702 for biogenic crust and 0.9473 for sand, with a difference of 0.0229. At 45 °C treatment, the average emissivity of biogenic crust and sand is 0.9725 and 0.9543 separately, with a difference of 0.0182. In the sand dune region, LST is usually between 35 °C and 50 °C at noon. Therefore, it can be concluded that the surface emissivity is about 0.97 for biogenic crust and about 0.95 for sand, with about 0.02 different.

Soils may have great difference of emissivity due to the difference of their properties. Many studies reported that bare silt soil has an emissivity of 0.96-0.98, sandy soil 0.93-0.96 and vegetation 0.97-9.99 (Hausenbuiller, 1978; Bruestets, 1982; Ladeb and Stoll, 1991a and 1991b, Humes et al., 1994). Biogenic crust in the region is actually composed of silt, fine sand, chlorophyll-contained microphytes, fungus and some debris of plants such as annuals. The grain-size of the sand in the region is also very fine (about 0.1-0.5mm). The color of sand is yellow. The mineral component of sand is mainly the dust and rocky particles from the North Africa desert. The quartz percentage of the sand is less than 50% in weight (Gerson et al., 1985). According to the soil properties of the region and compared with the reported soil emissivity in literature, it can be said that the results of the experiments are accep Table.

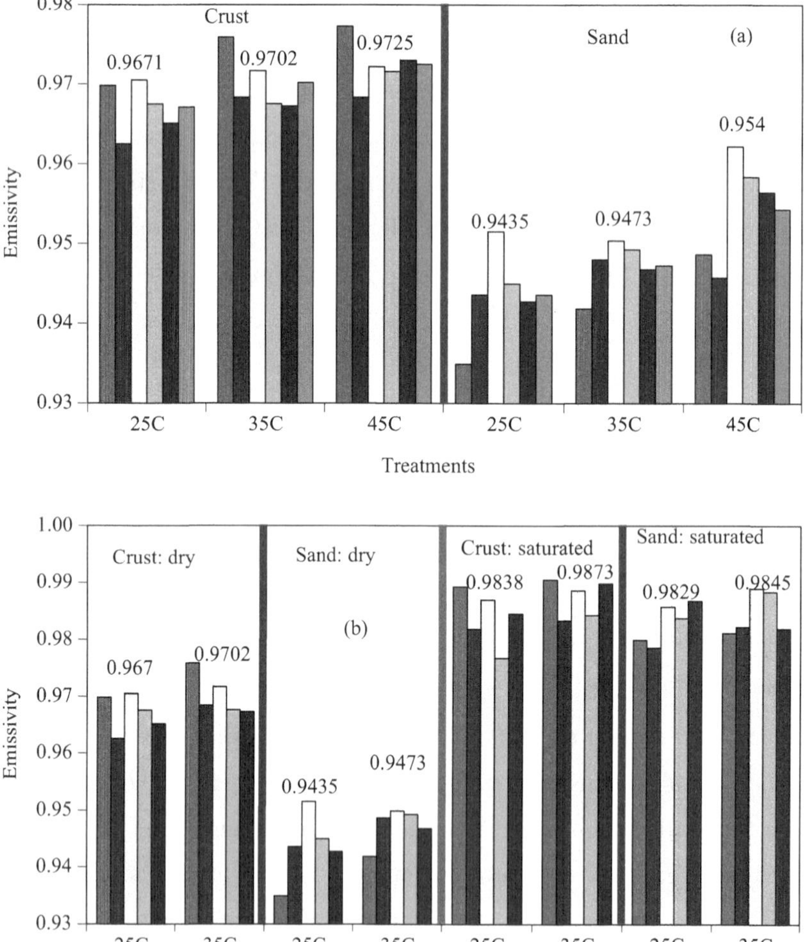

Figure 5.1 Emissivity of biogenic crust and active sand samples, illustrating (a) its variation in various temperature treatments and (b) its comparison with dry and saturation conditions. The number in the graph is the average emissivity of the samples from the treatment.

Another two features can also be seen from this change of soil emissivity in the experiments. Firstly, the emissivity of the samples increases gradually with temperature. For biogenic crust, the emissivity increases from 0.9671 at 25°C to 0.9725 at 45°C. From sand it changes from 0.9435 to 0.9543. The weak increase feature of the soil emissivity with temperature has been mentioned in many studies (Brutsaert, 1982; Hillel, 1980) and the results correspond very well with the feature. Secondly, the emissivity difference between biogenic crust and sand decreases with temperature. The difference is 0.0236 at 25°C and decreases to 0.0182 at 45°C. Therefore, one can derive that the emissivity of sand is more sensible to temperature change than biogenic crust. This probably is because that emissivity increase in temperature is not linear, but exponential (Brutsaert, 1982).

Table 5.1 Computation of emissivity based on laboratory experimental data

Samples	Treatment at 25°C			Treatment at 45°C		
	Temperature (K)		Emissivity	Temperature (K)		Emissivity
	Measured	Calibrated		Measured	Calibrated	
Reference	297.9306	298.1600	1.00000	318.4227	318.1600	1.00000
Crust 1	295.6622	295.8854	0.96983	316.5995	316.3368	0.97727
Crust 2	295.1044	295.3276	0.96254	315.8766	315.6139	0.96837
Crust 3	295.7156	295.9388	0.97053	316.1875	315.9848	0.97219
Crust 4	295.4868	295.7100	0.96754	316.1371	315.8744	0.97157
Crust 5	295.3035	295.5267	0.96514	316.2540	315.9913	0.97301
Average	295.4545	295.6777	0.96712	316.2110	315.9482	0.97248
Sand 1	292.9589	293.1821	0.93487	314.2584	313.9957	0.94866
Sand 2	293.6369	293.8601	0.94355	314.0169	313.7542	0.94575
Sand 3	294.2552	294.4784	0.95152	315.3716	315.4492	0.96219
Sand 4	293.7478	293.9710	0.94496	315.0606	315.2079	0.95840
Sand 5	293.5718	293.7950	0.94271	314.8972	314.7345	0.95641
Average	293.6341	293.8573	0.94353	314.7210	314.4582	0.95428

Note: The calibration coefficient is 0.2232 for 25°C and -0.2627 for 45°C.

Water is the most important factor shaping soil properties. The above determination of soil emissivity is based on dry samples corresponding to the arid environment of the region. However, in wet cool season, the surface of the region is not absolutely dry but contains some moisture. In order to compare the impact of soil water content on its emissivity change, the experiments have also been done for saturated samples. Considered the possible temperature change in wet season, only two treatments were conducted. The results are shown in Figure 5.1b, which compares the emissivity of the soil samples in dry and saturated conditions.

As indicated in Figure 5.1b, soil water eradicates the difference of possible emissivity between biogenic crust and sand. For 25°C treatment, the average emissivity at saturation is 0.9838 for biogenic crust and 0.9829 for sand. Statistically, there is no difference between them. The same situation can also be seen for 35°C treatment when biogenic crust has an average emissivity of 0.9873 and sand 0.9845. Actually, water has an emissivity of 0.99 (Lillesand and Kiefer, 1987; Sabins, 1986; Hillel, 1971). When soil moisture increases, thermal properties of the soil will bias from the pure soil ones in dry condition to the water ones at the extreme pole (Marshall and Holmes, 1979). At saturation, soil water content reaches the maximal level. When soil porosity is high such as the case in sand and the soft biogenic crust, the thermal properties of soil water at saturation prevails over those of soil in dry condition. And this is why the emissivity of biogenic crust and sand tends to be identical at saturation.

Conclusion of emissivity determination

Ground emissivity plays an important role in thermal emittance of the ground surface. Using emissivity, the observed radiant temperature can be calibrated into true or kinetic surface temperature through Stefan-Boltzmann law about radiation. In the study, the observed temperature from both ground truth measurement and remote sensing images is radiant temperature. Thus, emissivity is necessary for the examination of KST change of the region. Experiments based on the samples taken from the field have been carried out for determining emissivity of the two most important surface patterns.

According to the experiments and analysis, it can be concluded that the average emissivity of biogenic crust and sand is separately 0.97 and 0.95 in dry condition. Both biogenic crust and sand samples behave a weak increase of emissivity with temperature. The sand seems to have a slightly greater increase pace than biogenic crust. Comparison has also been done to the emissivity of the samples in dry and saturation. The result indicates that the difference of emissivity between biogenic crust and sand samples in dry condition seems to disappear at saturation when soil water governs the thermal properties of the soil. At saturation, both biogenic crust and sand samples have high and close high emissivity of up to 0.9829-0.9838 for the treatment at 25°C and 0.9845-0.9873 at 35°C.

Even though the biogenic crust and sand totally account for above 85% of the ground surface of the region, the emissivity of other two patterns (vegetation and playa) is also required to be considered for determination of average emissivity on both sides of the region.

Shrub is the dominant vegetation form of the region. The emissivity of vegetation can be found from many studies because vegetation is very important in shaping the environmental quality of the Earth's surface. Iddso et al. (1969) determined the emissivity values of a number of different plant leaves and found the emissivity ranges from 0.957 to 0.995. Davies and Idso (1979) concluded that most canopy emissivity values were likely to lie between 0.96 and 0.98 and believed a value of 0.98 is suitable in most radiation balance calculation. Humes et al. (1994) summarized the reported emissivity of vegetation as follows: grass with partial cover 0.956, shrub with partial cover 0.976 and with complete cover 0.986. Labed and Stoll (1991a and 1991b) got the emissivity of grassland (≈15cm) 0.983 and brushes (≈100cm) 0.994 according to their field measurements. Price (1984) reported that the emissivity of crops in agricultural field is in general 0.96-0.98. Considering the arid environment of the region, I use an emissivity of 0.975 for the arid shrubs.

As indicated above, playa of the region is similar with silt/clay soil. And many studies reported that clay soil has an emissivity of about 0.96-0.97 (Kahle, 1980; Koorevaar et al., 1983). For the study, playa only accounts for several percentages of the ground, hence has little contribution in shaping the average LST change on both sides. Thus, an average emissivity of 0.965 will be used for playa in the following computation.

5.4 Ground truth measurements of surface temperature

Report about the ground truth measurements of surface temperature change in the region has not been seen in literature. I did the measurements with an ain at validating what was observed on remote sensing images about the LST change on both sides.

Remote sensing detects the spectral signals of the Earth from space. Different surfaces have different spectral characteristics, hence are observed as difference in remote sensing (Hord, 1986; Harrys, 1987). As to the sand dune region, the field investigation indicates that the ground can be viewed as composed of four main patterns: biogenic crust, sand, playa and vegetation. Therefore, it is natural to think that the LST difference observed by remote sensing on both sides of the region is directly results from the difference of their surface composition structure. Moreover, It is the fact that the Israeli side is mainly covered with biogenic crust and various shrubs while the Egyptian side with sand (Pinker and Karnieli, 1995; Karnieli, 1997). Thus, it can be logically concluded that there will be an obvious kinetic surface temperature (KST) difference among the main surface patterns of the region.

In order to validate this assumption, the actual or kinetic surface temperature change of the typical surface patterns has been directly measured for further modeling and analysis. Due to the difficulties of accessing the border region, the measurement activities are not very regular, but the results do provide evidences to support the above assumption that the obvious LST difference observed on both sides can also be found on the main surface patterns of the region.

The measurements were done using an IR thermometer operating in 8-14 μm. The output of the thermometer is radiant surface temperature (RST). The advantage of using IR thermometer for the measurements is that it has the ability of quickly (usually in about 2-3 seconds) responding to the thermal radiance of the target and the output is parallel to what was obtained in remote sensing. The output frequency of the thermometer can reach to one record per second. A datalogger and a storage module were used to record the output of the thermometer.

The measurements were conducted at the Nizzana Research Site on the Israeli side (see Figure 5.4). Besides, the security problem and the difficulty to access the region also limit my measurement frequency and other field investigation activities. Each time we went to the region, we can only stay in the field for a few hours. Under these limitations, I did my best to carry on the ground truth measurements of surface temperature.

Several measurements of surface temperature change had been conducted at different times and different seasons. A number of sampling sites representing the main surface patterns had been selected for the measurements. It is the fact that ground surface temperature changes from time to time and only one thermometer was available for the measurements. In order to have a comparable result of the measurements, the sampling sites were selected to be adjacent in a small area and the measurements were conducted in a period of about 30-45 minutes so that the effect of different measuring time can be minimized. As for the measurements at each sampling site, the thermometer was pointed to the ground at the height of about 1 m with a viewing angle closed to nadir. The actual measuring time for each site is about 15-20 seconds so that at least 10-15 data was recorded for that site. The recorded data were then downloaded into a computer for further analysis. The average temperature was computed for each sampling site.

5.5 Correction of the measured RST into KST

The output of IR radiometer is the radiant surface temperature, which is converted from the observed thermal infrared radiance. However, this radiance reaching the sensor of the IR radiaometer is not only from the ground emittance but also composed of the reflected part of the atmospheric and the surrounding emttiance reaching the surface. Thus the direct output of the IR radiometer is generally call the radiant surface temperature (RST). What we want to measure is the true surface temperature or the kinetic surface temperature (KST) that determines the amount of ground emittance. Therefore we have to carry on correction procedure to transfer the RST into the KST in the laboratory after the measurement was done in the field. The correction of RST measurements into KST is based on the Stefan-Boltzmann law about thermal radiance or Planck radiance function. Simulation results indicate that the two methods have a difference of -0.1°C at low temperature (15°C) and of -0.4°C at high temperature (55°C). Since our measurements were conducted with an IR thermometer operating in the wavelength of 8-14μm, we preferred to base the approximation of the Planck function in order to have a better accuracy. For a temperature T, we can compute the Planck radiance as

$$B(T) = \frac{c_1}{\lambda^5 (e^{c_2/(\lambda T)} - 1)} \tag{5.7}$$

where $B(T)$ is Planck radiance (Wm^{-2}) of the ground, c1 and c_2 are the first and the second spectral constants, and λ is effective wavelength. The IR thermometer measures the target's temperature through detecting its radiance intensity and converting the observed radiance into temperature with a blackbody emissivity (ε=1). The total thermal radiance reaching the IR thermometer I can be divided into three components: ground emittance Ig, the reflected downward atmospheric emittance I_{ar} and upward atmospheric emittance $I_a^{\uparrow}$. Mathematically, we have

$$I=I_g+I_{ar}+I_a^{\uparrow} \tag{5.8}$$

The term I_{ar} depends on the ground emissivity and the downward atmospheric emittance $I_a^{\downarrow}$, generally expressed as

$$I_{ar}=(1-\varepsilon)\, I_a^{\downarrow} \tag{5.9}$$

where $I_a^{\downarrow}$ is the downward atmospheric emittance, which is strongly depended on the effective atmospheric mean temperature and the atmospheric vapor content and its distribution in the profile. In remote sensing, the value can be approximated as

$$I_a^{\downarrow}=(1-\tau_\lambda)B_\lambda(T_a^{\downarrow}) \tag{5.10}$$

where τ_λ is atmospheric transmittance at wavelength λ and $B_\lambda(T_a^{\downarrow})$ is the Planck radiance function. Simulation indicates that the magnitude of $B_\lambda(T_a^{\downarrow})$ ranges from 3.588-4.236W m^{-2} at $T_a^{\downarrow}$=10-20°C for the wavelength 8-14 μm. Since the region is an arid region and atmospheric water vapor is low (usually <2 g cm^{-2}), the atmospheric transmittance is high. Assumed τ_λ=0.8, the $I_a^{\downarrow}$ is estimated to be only about 0.72-0.84W/m^2. Furthermore, the ground emissivity of the region is about 0.95-0.97, which consequently results in the magnitude of the term I_{ar} quite small compared with the term Ig. Numerically, I_{ar} is estimated to be about 0.6%-0.8% of Ig in our region. The small portion of I_{ar} in comparison with I_g makes the simplification of computing KST from the measured RST possible. Moreover, the distance between the ground and the IR thermometer is only about 1 meter, which makes the item $I_a^{\uparrow}$ too small to be considered. Thus, we have

$$I=I_g+I_{ar}\approx 1.007 I_g \tag{5.11}$$

Substituting the total radiance observed by the IR thermometer $I=B(T_r)$ and the ground emittance $I_g=\varepsilon B(T_k)$ and using equation (5.7) for Planck radiances $B(T_r)$ and $B(T_k)$, we obtain the following experimental formula for approximating T_k from the observed RST in our region:

$$T_k = \frac{c_2}{\lambda \ln[1.007\varepsilon(e^{c_2/\lambda T_r} - 1) + 1]} \tag{5.12}$$

where T_r is the observed RST of the ground (K), T_k is the true KST of the ground (K), c_2 is the second spectral constant with c_2=1.43876869×10^{-2} m K, ε is ground emissivity, and λ is the effective mean wavelength of the IR thermometer with λ≈12.15μm Therefore, knowing the emissivity, the observed RST can be easily corrected into KST. The corrected results of the ground temperature measurements on surface temperature of the region are presented as follows.

5.6 Results and analysis of the surface temperature measurements

5.6.1 Surface temperature change in hot dry season

The climate of the region belongs to the typical Mediterranean one. It has two distinctive seasons: dry and wet. Dry season usually starts from May and ends in about October. The rest of the year is defined as wet season though the rain during the season is also very little. During the dry season, there is no rain and the sky is very clear. Under the baking of sunshine, the surface of the arid region is very hot at noon. June, July and August are the hottest months of the year.

Figure 5.2 represents the typical KST change of the region in hot dry season. Several features can be found from Figure 5.2a, which shows the measured result taken at noon on June 18, 1997. There is an obvious KST difference among the surface patterns. Biogenic crust has the highest KST and vegetation the lowest. The average KST of biogenic crust is 54.11°C and some sampling sites even reach up to 55°C. KST of playa is slightly higher than that of sand. Average KST of playa is also high up to 51.80°C and that of sand is 50.97 °C (Figure 5.2a).

The KST difference between biogenic crust and sand is very obvious and it is highly up to above 3.14°C at noon (Figure 5.2a). The KST of biogenic crust is also about 2.31°C higher than that of Playa. Another feature is the sharp KST difference between shrub canopy and its surrounding surface. According to the measurement, the difference is up to 10-15°C (Figure 5.2a). This implies that the transpiration from vegetation still controls its KST change even though desert shrubs have some special functions to reduce its water lose in hot summer (Danin, 1983; 1991).

Similar features of KST change can also be seen in Figure 5.2b. The KST

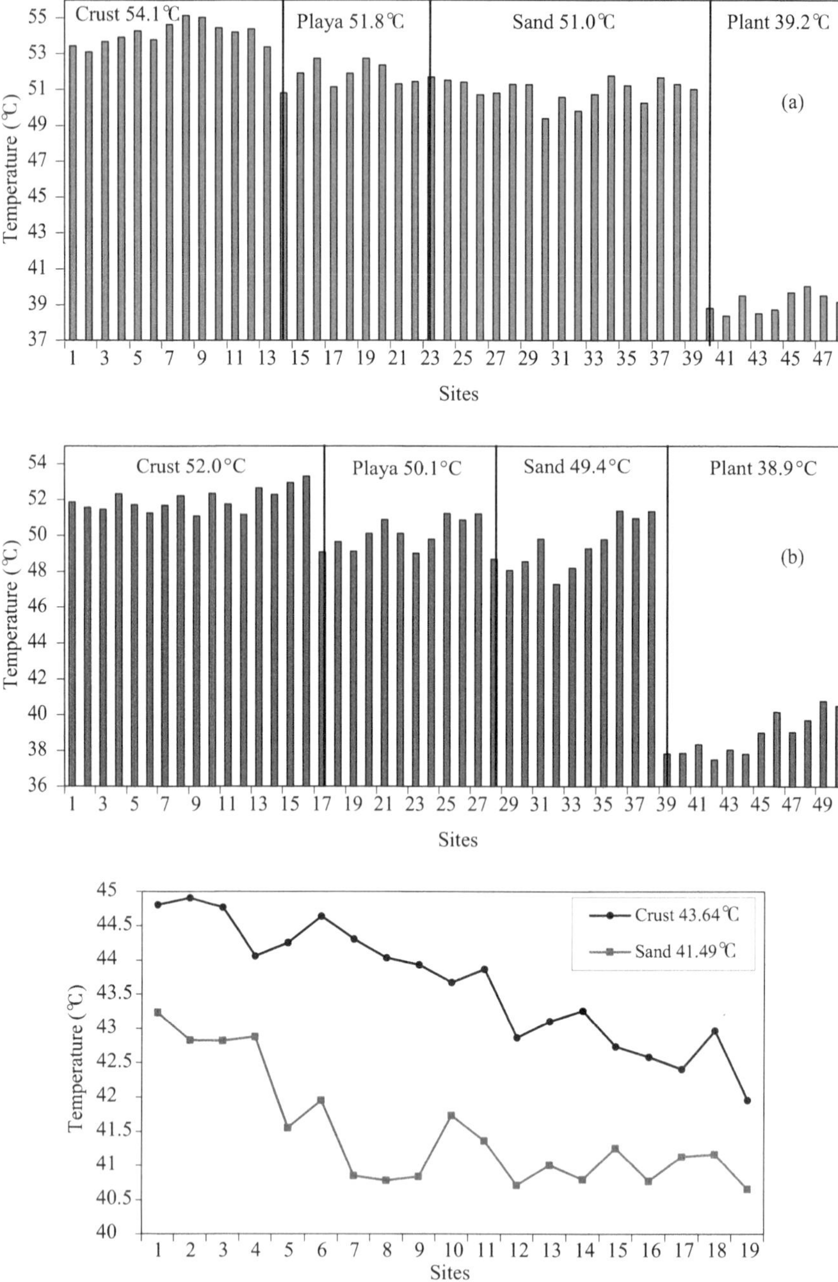

Figure 5.2 Surface temperature change of typical surface patterns in hot dry season, illustrating the measurements at (a) 13:32-14:05, June 18, 1997, (b) 11:30-12:15, August 1, 1998 and (c) 16:15-16:48, June 18, 1997. The number in the graph is the average KST of the patterns

difference among the four surface patterns is also very obvious. The average KST of biogenic crust is about 51.97°C, which is 2.54°C higher than that of sand and about 1.88°C playa and 13.09°C plant. Please note that the measurement was taken at the time a little bit ahead of the temperature peak at noon. According to the change trend of the measurement, the difference in the noon peak period may reach the level of above 3°C between biogenic crust and sand. A gradually ascending of KST change can be clearly seen during the measurement period. For biogenic crust, the KST ascends from about 51.5°C at the beginning to 53°C at the end of the measurement. The KST of sand, Playa and plant also shows some increase trend during the measurement (Figure 5.2b).

Special comparison of KST change between biogenic crust and sand is presented in Figure 5.2c, which represents the typical KST change of the two surface patterns in afternoon during hot summer. The ground surface was experiencing a cooling process due to the decrease of global radiation. For biogenic crust, KST decreases about 2.5°C within half an hour of the measurement. It seems that the cooling speed of sand is almost the same as biogenic crust (Figure 5.2c). Again, the temperature difference between the two surfaces is very obvious. On average, it is about 3.15°C, almost the same as it was at noon (Figure 5.2a).

5.6.2 Surface temperature change in cool wet season

Annual precipitation of the region is only about 95mm (Kidron, 1997) and most of the rainfall concentrates in one or two main raining months (Evenari et al., 1971). These characteristics make the other months of wet season not really wet. The measurement shown in Figure 5.3 represents the KST change of the region under very wet conditions. There was a strong raining process with total precipitation of 30.5mm from January 5 to 12, 1998. The KST shown in Figure 5.3a was measured after 2 days of the raining process. The measurement time was in the morning between 9:00 and 9:25.

Field investigation indicated that the ground was obviously wet. Some low places still had some water on the surface. After the rain, the microphytes on the crust surface became active. Several relatively dry sites were selected for the measurement of KST change.

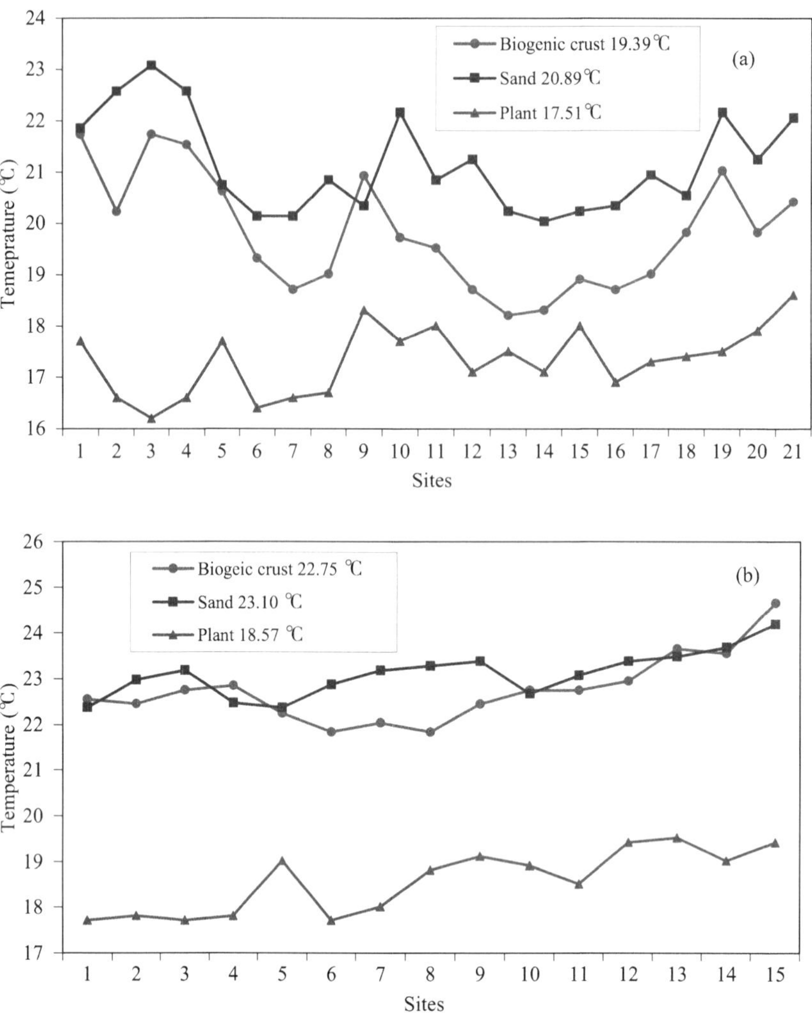

Figure 5.3 True surface temperature change of typical surface patterns in very wet cool season, illustrating the measurements at (a) 9:00-9:25, January 16, 1998 and (b) 10:00-10:35, January 25, 1998. The number in the graph is the average temperature of the measurement sites.

The measurement results (Figure 5.3a) show that the sand has KST of higher than biogenic crust. The average KST of sand is 20.89°C, which is 1.50°C greater than that of biogenic crust (19.39°C). Two reasons may lead to the existence of this phenomenon. The first is that the wet surface wipes the potential KST difference between the two surface patterns. Water is a critical factor that shapes the soil properties for heat transfer (Marshall and Holmes, 1979). In a short time after the rain, the soil under the surface is filled with a lot of water. The high content of soil water may have casignificant contribution to narrowing the difference of surface temperature change. Another reason is that the measurement was taken in the morning at about 9:00 when the surface temperature may differ slightly on the two surface patterns. In combination, these two factors result in the disappearance of possible surface temperature difference between sand and biogenic crust.

However, the canopy temperature of shrub still remains lower than that of sand or biogenic crust (Figure 5.2a). The average of the canopy temperature is 17.51°C, which is about 3.38°C lower than the KST of sand.

There was no rainfall between January 16 and 25, 1998. Therefore, the results shown in Figure 5.3b represent the KST change of the region after about two weeks of a strong raining process. Field observation can see that the ground is much drier than what was seen on January 16, 1998. According to the results in Figure 5.3b, the KST difference between the two main surfaces is narrowing. The KST during the measuring time shows a trend of gradual ascent, climbing from about 22.5°C to 25°C. Average KST is 22.75°C for biogenic crust and 23.10°C for sand. The difference is only 0.35°C, not statistically significant.

The vegetation canopy temperature still remains much lower than KST of biogenic crust and sand (Figure 5.3b). The average canopy temperature is about 18.57°C, which is about 4.2-4.5°C lower than the KST of sand and biogenic crust.

5.6.3 Surface temperature change in late wet season

Except for the above two extremely dry and extremely wet cases, several measurements were also conducted in various seasons. Figure 5.4 shows the KST change on March 19 and 26, 1997. Both the measurements were carried on in the late morning. Therefore, similar change trend can be found in both cases. For all four surface patterns, a slight ascent of KST change can be seen in both Figure 5.4a and 4.4b. Another feature is that there is an obvious KST difference among the four surface patterns. On March 19, the average KST of biogenic crust was 35.64°C and that of sand was 33.41°C, with a difference of about 2.23°C. On March 26, the KST difference was about 2.74°C.

The results in Figure 5.4 also indicate that the average KST of Playa is higher than that of sand but lower than that of biogenic crust. According to the measurement, the sampling sites of playa had an average KST of 34.36°C on March 19, 1997, which was about 1.28°C lower than biogenic crust and 0.95°C higher than sand. Figure 5.4b indicates that the KST difference between playa and biogenic crust was about 2°C and between Playa and sand about 0.84°C on March 26, 1997. This evidence supports the conclusion that under dry condition playa in the region has higher KST than sand but lower KST than biogenic crust. The conclusion is identical to what was getting from the measurement in hot dry season (Figure 5.2).

March and April are the most important growing months of many perennials (shrubs) and annuals in the arid region (Danin, 1983). The canopy temperature of shrubs appears to be similar in the growing season. On March 19 1997, the average KST of shrubs was about 25.64°C and on March 26 about 26.27°C, with only a difference of 0.53°C. However, the difference of shrub canopy temperature and the KST of other three main surface patterns is still very sharp. From Figure 5.4a one can see that at about noon on March 19, 1997 the canopy temperature of shrub was about 10°C lower than the KST of its surrounding biogenic crust and about 8°C lower than that of sand surface. Figure 5.4b depicts a similar KST difference between a shrub and biogenic crust and sand surface. On March 26, shrub canopy temperature was about 8.3°C lower than the KST of biogenic crust and 5.4°C lower than that of sand surface.

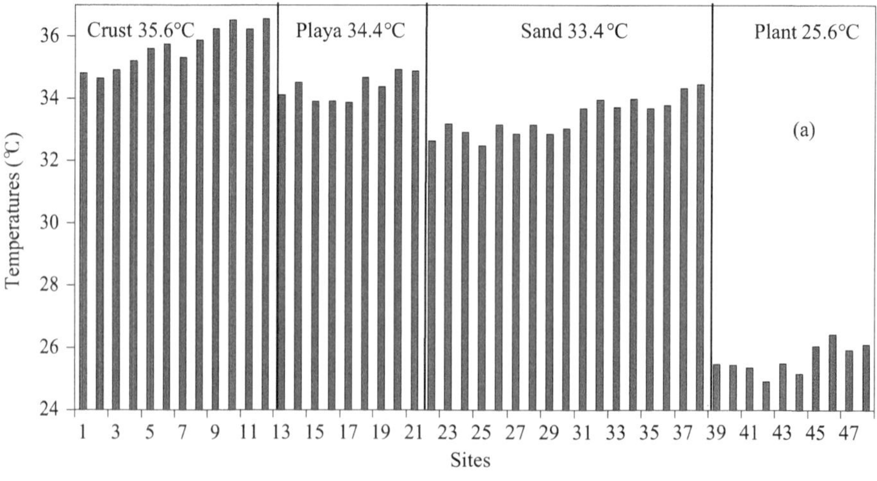

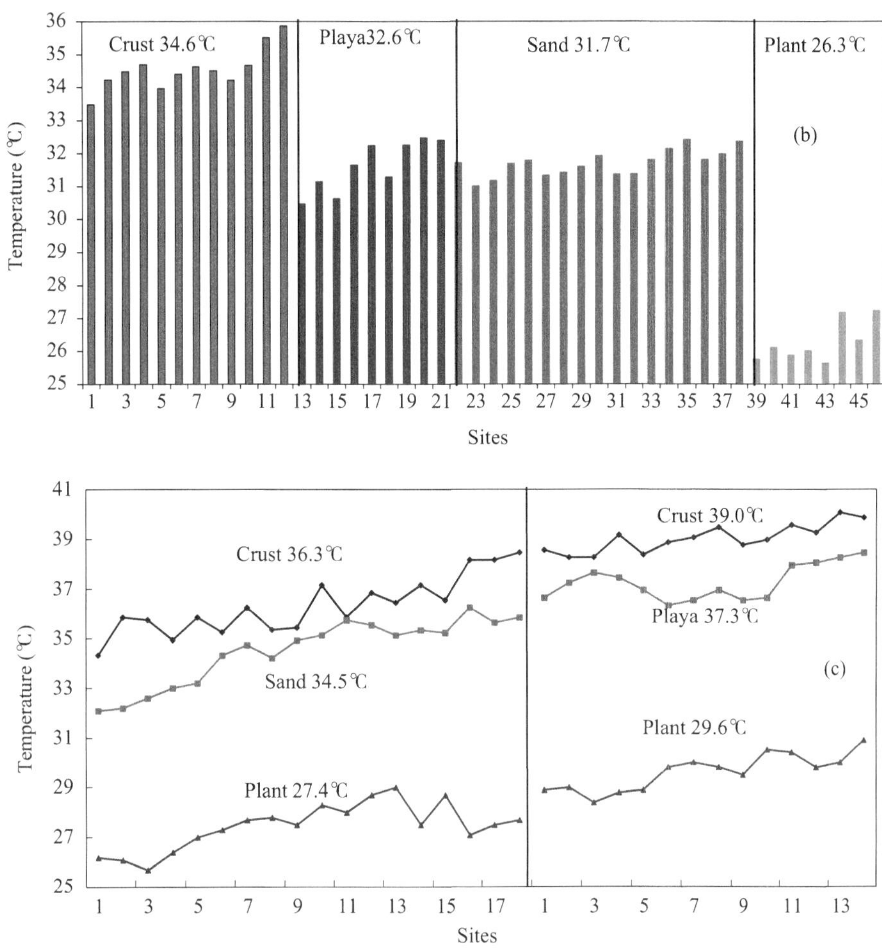

Figure 5.4 Surface temperature change of typical surface patterns in late wet season, illustrating the measurements at (a) 11:20-12:00, March 19, 1998, (b) 11:16-12:04, March 26, 1998 and (c) 10:30-11:11, April 10, 1998. The number in the graph is the average KST of the patterns.

Figure 5.4c shows another measurement for the comparison of KST difference among the four patterns in the late wet season. The measurement was carried on April 10, 1998 when the annuals were blooming. The left part of Figure 5.8 compares the KST change of biogenic crust with sand and shrubs and the right part compares the KST change of biogenic crust with playa and plant. Because the measurement was taken in the morning when the ground was experiencing a heating process, the KST appears in a trend of gradually ascending. The left part of Figure 5.4c indicates the different pace of KST ascent of biogenic crust, sand and plant. The KST of biogenic crust climbs from about 33°C to 35°C during the measuring period, with an average of 36.28°C. For the sand surface, KST increases about 4°C (from about 28°C to 32°C) during this period. A relatively weak increase can be found in the canopy temperature of shrubs, which lifts about 2°C (from about 24 to 26°C) within the period. Similarly, a sharp KST difference can also be seen in the left part of Figure 5.4c. The difference is about 1.7°C between biogenic crust and sand and about 8.8°C between biogenic crust and plant.

The right part of Figure 5.4c was taken to another sampling place for the comparison of KST differences on biogenic crust, playa and plant. The average KST of biogenic crust is about 36.67°C, which is about 2.3°C higher than that of playa and 9.4°C higher than that of plants. Comparing the behaviors of KST change of biogenic crust with that of sand and playa, one can see that the KST difference between biogenic crust and sand is slightly greater than it is between biogenic crust and playa. This evidence further strengthens the above conclusion that playa has slightly higher KST than sand.

5.6.4 Surface temperature change in early dry season

KST change of the four patterns in the early dry season can be found in Figure 5.5, which shows the measurement results taken on May 7, 1997 when it had been no rain for more than one month and the annuals were dead. At that time, the surface returns to a very dry state. The measurement shown in Figure 5.5a was taken in the early afternoon when the surface just passed its heating peak and started the cooling process. The temperature of the sampling sites shown in Figure 5.5a appears a trend of slightly descending. For biogenic crust, the KST descends from about 50°C to 48°C during the measurement, with a drop of about 2°C. The descending process can also be seen in the cases of sand, playa and plant, with only difference in their drop pace.

The obvious difference of KST change in various surface patterns is the same as those described in the above paragraphs. During the measurement, the sampling sites have an average KST of 48.69°C for biogenic crust, 46.47°C for playa, 45.31°C for sand and 35.93°C for plant. Therefore, the KST difference is up to about 3.38°C between biogenic crust and sand, about 2.12 °C between biogenic crust and playa and up to 12.76°C between biogenic crust and plant.

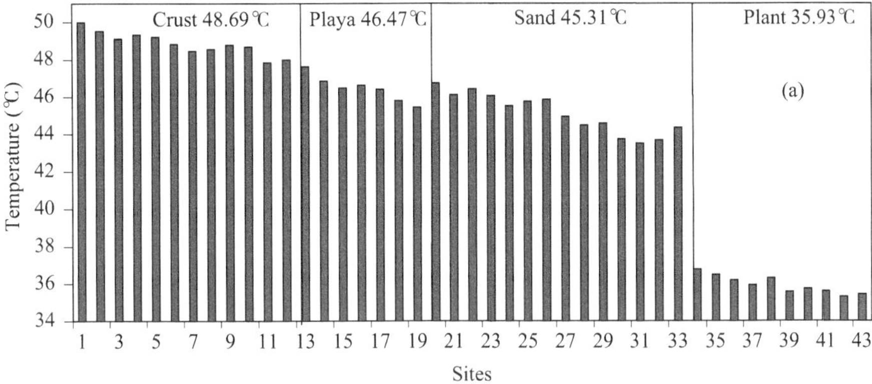

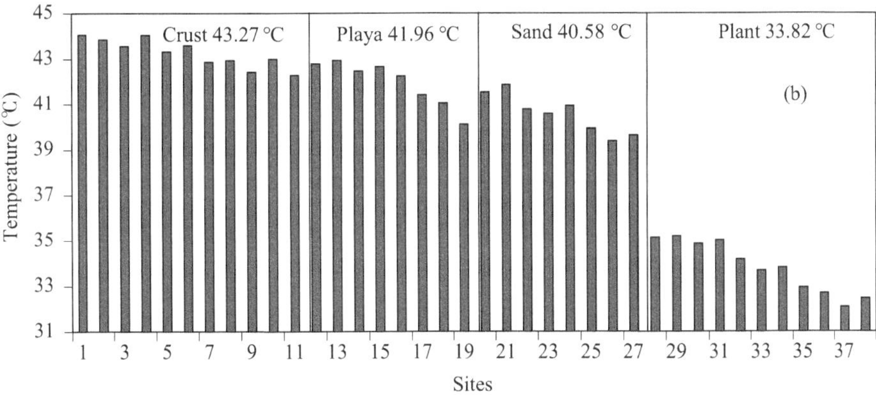

Figure 5.5 True surface temperature change of typical surface patterns in the early dry season, illustrating the measurements at (a) 14:15-14:52, May 7, 1997 and (b) 15:39-16:05, May 7, 1997. The number in the graph is the average temperature of the measurement sites.

The cooling process of the surface patterns in the afternoon of May 7, 1997 is shown in Figure 5.5b. Similarly, a gradual KST decent can be clearly distinguished for all surface patterns. The KST difference is also very obvious among the four patterns. Average KST of biogenic crust is about 43.27°C, which is about 1.31°C higher than that of playa and 2.69°C higher than that of sand respectively. The average shrub canopy temperature is about 9.45°C lower than the average KST of biogenic crust. Considered the cooling speed in afternoon, the relative lower KST difference in Figure 5.5b than in Figure 5.5a is reasonable and understandable.

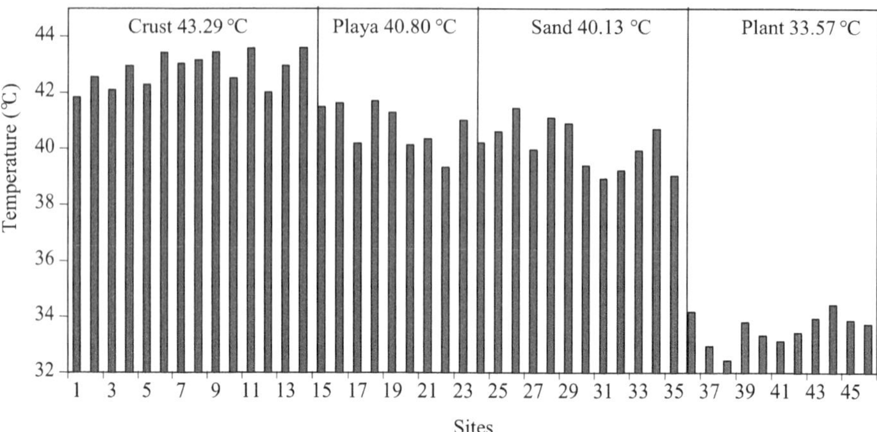

Figure 5.6 True surface temperature change of typical surface patterns in late dry season, illustrating the measurements at 14:19-14:55, October 6, 1997. The number in the graph is the average temperature of the measurement sites.

5.6.5 Surface temperature change in late dry season

The region enters wet season in late October or early November of the year, depended on the time of first raining (Evenari et al., 1971). The situation in 1997 was that the first rain did not come until October 16 (0.1mm) and 17 (8.1mm). Thus, when the measurement was taken on October 6, the region still remained in dry season even the temperature has dropped down due to the solar angle decrease. The measurement was carried on from 14:19 to 14:55, which was at about temperature peak time. The change trend of KST is very weak during the measurement (Figure 5.6). Average KST of biogenic crust is about 43.3°C, which is also obviously higher than that of playa, sand and plant. According to the measurement, biogenic crust has a KST of about 2.5°C higher than playa, about 3.2°C higher than sand and about 9.7°C higher than shrub canopy. This KST difference again supports the argument that an obvious LST difference exists between biogenic crust and sand, which dominate the majority of ground surface in the region.

5.6.6 Conclusion of ground truth measurements

During the period 1997-1998 several ground truth measurements had been conducted in various seasons at different times of the day. The purpose of the measurements is to validate what was observed in remote sensing analysis about thermal variation of the region. Though the temperature is a changeable phenomenon, the measurements based on a number of sampling sites in a relatively short time can still reveal some general trends of its variation.

The results of ground truth measurements confirm what was expected, according to the observed LST change on both sides in remote sensing analysis. Though the measurements are based on a few sampling sites, an obvious difference of surface temperature does exist between biogenic crust and sand in most days of the year. In dry summer at about noon, biogenic crust has higher KST than sand and the KST difference between the two main surfaces is high up to above 3°C. During the morning and afternoon, the difference is also obvious. In early and late dry seasons, the difference is still high up to above 2.5°C at about noon. The difference reverses only in a few days after heavy rain when biogenic crust is very wet or at a level of high water content while sand surface is relatively drier. Under this extreme condition, sand has KST of about 1°C higher than biogenic crust. The lower KST of biogenic crust is because that soil water content strongly shapes its thermal properties that govern KST change. And raining means soil water content increases. However, the reverse KST difference between the two main surfaces becomes very weak after about two weeks of strong rain (Figure 5.3b). Therefore, in wet season, biogenic crust has higher KST than sand in most of the days except the raining days and a few day after the raining process. Figure 5.3 is a typical example illustrating this situation.

The plant has the lowest canopy temperature in all seasons. Usually in hot dry summer, the difference of shrub canopy temperature from biogenic crust KST may be up to 13°C at noon, as indicated in Figure 5.2a. In wet season when the KST difference between biogenic crust and sand surface reverses due to the wetted surface of biogenic crust, shrub canopy temperature still remains about 3°C lower than KST of sand, as indicated in Figure 5.3a.

The ground truth measurements can also present the possible levels to which the KST of the main surface patterns may reach in various seasons. In dry summer, KST of biogenic crust may reach up to 52-55°C at about noon. The KST of sand and playa may be up to 50-53°C and 51-53°C, respectively. Shrub canopy also has high temperature at noon in the hot dry season. In wet winter, the KST of biogenic crust and sand may reduce to the level of below 20°C in a week after a strong raining process and the shrub canopy temperature may drop to about 16-17°C under this condition. However, if there was no rain in about two weeks, the KST of biogenic crust and sand will recover to the level of above 22-25°C. Usually the KST of biogenic crust is up to 38-40°C at noon in late wet season. The sand surface and shrub canopy in the season may have KST of 35-37°C and 27-30°C respectively. Knowing of these changes is important for simulating the mixed KST difference on both sides of the region, because it is impossible to cross the border for any measurements on the Egyptian side.

6 Surface composition and its impact on LST change

Considered the lithological and geomorphological similarity on both sides, the sharp contrast of LST on both sides of the Israel-Egypt border should be caused by the different composition structure of surface patterns. Thus, the estimation of land cover structure on the both sides is necessary to understand the generation of this abnormal thermal phenomenon. Analysis of the ground truth measurements of surface temperature change in the four typical surface patterns in previous chapter indicates that the main surface patterns really have an obvious difference of surface temperature change in various seasons. However, the true surface temperature change based on a mixed surface assumption has not been known for the surface composition structure, which is the focus of this chapter.

6.1 Field observation of surface composition structure

As indicated above, the region can be viewed as composed of four main surface patterns: biogenic crust, sand, vegetation and Playa. It is the difference of surface composition that leads to the sharp contrast of LST change on both sides of the region. Therefore, knowing the quantitative composition of the ground surface of the region is extremely important for an explanation of the interesting LST anomaly in the study. However, the accurate percentage of each surface pattern has not been reported in the existed studies about the region. Though many studies (Ottermann, 1974; 1977; 1981; Tsoar and Moller, 1986; Karnieli, 1997) relating to the region asserted denser vegetation cover on the Israeli side, only Kidron and Yair (1997) Karnieli and Tsoar (1995) and Otterman and Robinove (1982) mentioned the value of the cover. Kidron and Yair (1997) arbitrarily guessed 10%-40% of vegetation cover, Karnieli and Tsoar (1995) lower than 30% in the region and Otterman and Robinove (1982) about 20% on the Israeli side. Details about plant communities at Nizzana Research Site was investigated in Tielbörger (1997), who also reported the percentage cover of each perennial plant communities at the Site. The percentage of plant cover is different from community to community, ranging from 1.9% in Anabasis articulata to 32.1% in Noaea mucronata-Artemisia monosperma (Tielbörger, 1997). The exact percentage of other surface patterns such as biogenic crust and sand has not been reported yet. In order to get these data, three methods have been employed for the estimation of the surface structure: field observation and statistics, measuring on aerial photographs, and measuring vegetation cover on Landsat TM and SPOT images.

6.1.1 Field observation for the measurement

Field observation for the measurement was done on March 26, 1998 at Nizzana Research Site (see Figure 6.4). The primary purpose of the measurement is to determinate the vegetation percentage on the Israeli side. Five plots with various landscapes had been selected for the measurement. The distribution of the five plots was as follows: two plots in the interdune, one at the foot of dune slope, one in the middle and one on the top dune. The surface of the sampled plots was mainly constituted of biogenic crust and vegetation (mainly shrubs) except the plot on top dune, which had the surface structure as sand and vegetation.

Firstly, the size of each plot was measured and the total number of shrubs was counted. Then, a number of shrubs were randomly selected to measure their diameter for statistical analysis because measuring all the shrubs in the plot was a time consuming burden. Finally, the average shrub diameter was calculated for estimating vegetation area. The vegetation cover rate is given by dividing the vegetation area to the total plot area.

Some small shrubs do not fully cover the land. Only those big shrubs (usually with diameter >1.5m) are able to have a full cover with their thick dense canopy. This makes the measurement difficult. However, only those shrubs with moderate size (diameter above 0.5m) were counted as vegetation and those with small diameter (less than 0.5) were neglected. This sampling rule can compensate some errors that might arise in the measurement. Because shrubs were perpetual, the measurement also represents the vegetation rate in dry season.

The sampling process and results are shown in Table 6.1 and 6.2 respectively. Table 6.2 indicates that there is no difference of vegetation cover between interdune and dune slope. However, it seems that the distribution of vegetation is far from even in the region. The vegetation cover rate tends to be different between top dune and dune slope or interdune. According to the measurement, the average vegetation cover rate is about 15% on the Israeli side, with interdune and dune slope about 13%-19% and top dune about 11%. The biogenic crust surface accounts for above 2/3 while the sand for about 15%.

Table 6.1 Sampling process for estimating vegetation cover at Nizzana Research Site

Plot	Location	Total shrub number	Shrub number. sampled	Average diameter (m)	Average area (m^2)	Shrub area (m^2)
1	Interdune	198	27	1.062	0.8849	175.218
2	Interdune	121	28	1.266	1.2588	152.315
3	Bottom slope	168	30	1.053	0.8709	146.304
4	Middle slope	42	15	1.087	0.9273	44.511
5	Top dune	37	9	0.754	0.4470	75.909
Total of the samples		566	109	1.094	0.9395	594.257

Note: Plot 5 has 7 big shrubs with total area of 62.5 m^2. Thus, the shrub area of this plot is equal to the area of 30 normal shrubs plus the area of the big shrubs.

6.1.2 Statistical estimation from the field observation

The above results may not be very accurate because the plots in the interdune seem to be sampled in a small proportion when compared to the sand plot. At the same time, playa (also named physical crust), one of the important surface patterns, is not counted in the measurement. Therefore, calibration needs to be done by taking a cross-section measurement, as indicated in Figure 6.1.

Table 6.2 Calculating vegetation rate on Nizzana Research Site

Plot	Location	Area (m^2)			Percentage	
		Total	Shrub	B. crust	Shrub	B. crust
1	Interdune	1330	175.218	1154.782	13.174	86.826
2	Interdune	780	152.315	627.685	19.527	80.473
3	Bottom slope	740	146.304	593.696	19.770	80.230
4	Middle slope	330	44.511	285.489	14.848	85.152
5	Top dune	671	75.909	595.091 (S)	11.313	88.687 (S)
Total of the samples		3851	594.257	2661.652	15.431	69.116

Note: The value with (S) refers to sand.

The total length of the cross-section under study is 498m, of which 64m is the north interdune, 135m the north dune, 148m the south interdune and 151m the south dune. Except shrubs, the interdunes are mainly covered with biogenic crust and Playa, the tope dunes with sand and the slopes with the biogenic crust. Playa only accounts for a small proportion in the interdunes (3m in the northern one and 12m in the southern one).

Therefore, biogenic crust with shrubs accounts for (197+221)/498=83.936% of the total, while sand with shrubs for 13.052% and playa for 3.012%. Playa plot is not sampled because there are almost no shrubs on it. So, the calibration should also not consider its effect. Under this consideration, the ratio of biogenic crust with shrubs to the total is 418/(498-15)=86.542% and that of sand with shrubs 13.458%.

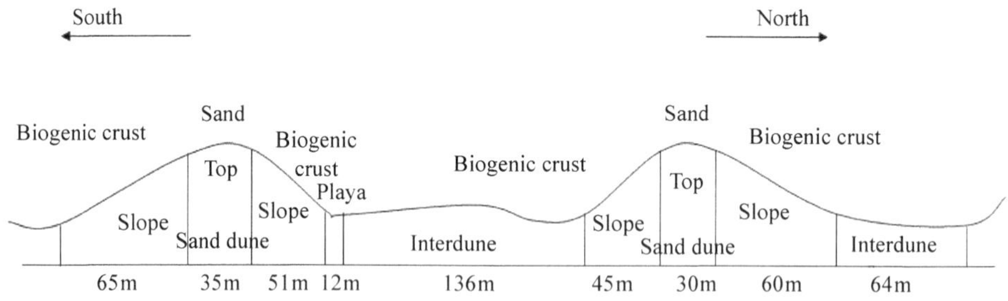

Figure 6.1 The cross-section of two sand dunes in the region.

Taking into account the proportion of the two surface patterns, the weighted area (m^2) of the each component is as follows: biogenic crust with shrubs (2110+1070)×86.542%/82.576% =3332.73m^2, of which shrub 3332.731×16.30%= 543.24m^2 and crust 2789.49m^2; sand with shrubs 518.25m^2, of which shrub 58.63m^2 and sand 459.62m^2. Based on this calibration, the cover rate of each pattern is re-calculated as biogenic crust 69.42%, sand 11.94%, vegetation 15.63% and playa 3.01%.

Compared with the directly sampling results, the vegetation rate remains stable. The biogenic crust cover rate increases to about 72%, while the sand decreases to about 12%. The biogenic crust percentage includes the effects of playa because Playa is always located in interdune where biogenic crust prevails. If this effect is considered, the biogenic crust cover rate will be 69.42% and the playa 3.01%. These results seem to be more reasonable according to the field observation. Before calibration, sand has a higher fraction than vegetation. After calibration, these two surface patterns seem to have equal percentage. Field observation indicates that sand surface can only be found on top dune with various widths ranging from 10-40m. Shrubs can be found in the interdune, dune slope and even top. Even though these shrubs appear as scattering distribution in the region, the cover rate should not be less than sand.

The measurement results subject to the sampling limitation because only five plots with 3851 m^2 were analyzed, which is far smaller in comparison to the total region with several hundred squared kilometers. Another factor is the heterogeneous distribution of shrubs in the region. The sampling site locates in the southern part of the region, where top dune sand seems to be more and vegetation sparser than in the northern and central part of the region. In order to improve the estimation, an aerial photograph with high resolution was used for the measurement.

6.2 Measuring on an aerial photograph of the region

The aerial photograph for the measurement (Figure 6.2) covers both sides of the region. The part of Egyptian side is about 0.8 km in width and that of the Israeli side about 3 km. The location of the photograph is in the southern part of the region and it covers five sand dunes. The spatial resolution is high enough to clearly distinguish the shrub canopies from their surrounding background. The photograph has three channels in visible range.

The method used for measuring vegetation cover age rate of the image includes making subset image for both sides and taking measurement on the subsets. The spectral characteristics of the photograph indicate that shrubs appear as low DN pixels and sand as high DN pixels in all the three channels. Playa also has high DN value. The DN value of biogenic crust is between shrub and sand. A histogram of the pixels of the image indicates that there is a valley of the pixel distribution in the DN range 60-70 in both channels 1 and 2.

Usually, the measurement of the surface pattern on an image is done by using classification function of imagining analysis. I try to classify the two subsets imagines into the four patterns. My first choice is to use supervise classification with training points in the four patterns. I tried several times, but the results were not as good as expected. There were always some misclassifications among the four patterns when using all channels. The reason maybe that it is difficult to distinguish playa, and biogenic crust from sand pattern due to their close DN value. Unsupervised classification is also unable to result in a satisfied result. Thus, I decide to classify firstly shrubs from one channel, and then use interpretation method to identify the other patterns.

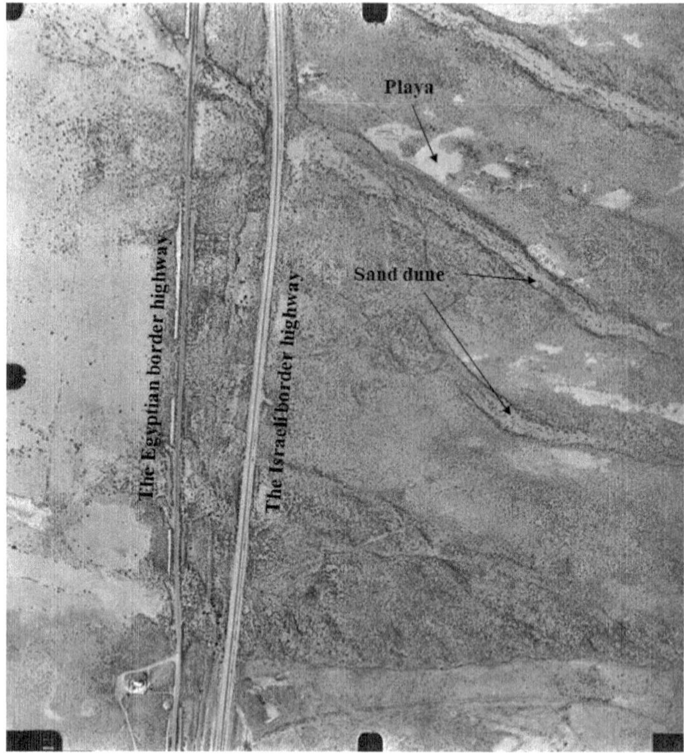

Figure 6.2 The aerial photograph used for the measurement of surface composition structure. Two border highways are clearly seen. Difference on both sides is very obvious. The Israeli side is darker than the Egyptian one due to more vegetation and biogenic crust. The dark dots are mainly shrub canopies and vegetation debris, the bright plots are playa, the yellow color on the Egyptian side is bare sand and the gray background on the Israeli side is biogenic crust.

Because vegetation has low DN value, I use channel 1 to test several shrub canopies. Most pixels of the tested shrub canopies have a DN value less than 65 and only few pixels on the edge have DN value higher than 65 but less than 80. Statistical analysis of the photograph indicates that the Israeli subset has total 3698857 pixels and Egyptian one 8896446. Pixels with DN $\leqslant 60$ account for 17.02% on the Israeli subset and 6.20% on the Egyptian one. Pixels with DN $\leqslant 65$ account for 18.29% on the

Israeli subset and 6.71% on the Egyptian one. Accordingly, it seems that vegetation cover rate on the Israeli subset is about 17% and on the Egyptian one about 6.0%. This obvious difference in vegetation coverage has significant contribution to the sharp contrast of spectral reflectance in remote sensing imagines as mentioned in Otterman (1974; 1977; 1981), Otterman et al. (1975), Warren and Harrison (1984), Tsoar and Møller (1986), Karnieli and Tsoar (1995) and Tsoar and Karnieli (1996).

Interpretation of the photograph (Figure 6.2) can clearly distinguish the distribution of other three surface patterns due to their special feature of spatial concentration. It is easy to draw a line between patterns according to their spatial location. Thus, in order to calculate their cover rate, I first interpret the polygons for these surface patterns on the photograph and then measure the area of these polygons. Some plots are not pure surface patterns but mixtures with some shrubs on them. This is especially true on biogenic crust on which various shrubs are rampant. The top dune sand also has some shrubs on the surface. This makes the measurement complicated because the vegetation cover has to be subtracted from the area of the polygons. The detailed measurements on the two subsets are given as follows.

6.2.1 Measuring on the Egyptian subset

On the Egyptian subset, the ground is dominant with sand. So I decide to measure the relatively small plots of biogenic crust mixed with shrubs and playa with shrubs. The area of sand can be gained by subtracting vegetation, biogenic crust and Playa from the total. Seven biogenic crust plots mixed with shrubs and two playa plots with shrubs are identified and sampled for the measurement on the subset. The results are shown in Table 6.3. Accordingly, the surface composition structure on the Egyptian subset can be estimated as follows. On average, vegetation accounts for 6.40% of the total, biogenic crust for 27.27%, playa for 3.29% and sand for 63.04%.

Table 6.3 also indicates that the Egyptian side has great differences of vegetation cover on the biogenic crust surface. Some plots have vegetation cover rate of high up to about 16%, while some only have about 5%. The two playa plots also have obvious difference of vegetation cover. The average vegetation cover rate on biogenic crust is about 12% while the percentage on playa is only 1.4%. In order to compare vegetation cover on the sand surface, five plots were sampled for the measurement, which gives the result as shown in Table 6.4. Therefore, the average vegetation cover rate on different mixed surface patterns can be concluded as 3.15% on sand, 1.44% on playa and 12.15% on biogenic crust.

Table 6.3 Measurement of vegetation cover on the Egyptian subset

Patterns	Plot	Area (ha)	V %	BC %	BC area (ha.)	BC pixels
BC+V	1	62.2094	11.8498	88.1502	54.8377	548377
BC+V	2	33.1814	4.7101	95.2899	31.6185	316185
BC+V	3	5.02474	5.9121	94.0879	4.7277	47577
BC+V	4	95.7241	15.9148	84.0852	80.4898	804898
BC+V	5	15.6766	14.8353	85.1647	13.3509	133509
BC+V	6	53.2893	10.7823	89.2172	47.5432	475432
BC+V	7	11.0809	9.2356	90.7644	10.0575	100575
Total		276.1864	12.1516	87.8484	242.6254	2426254
Playa+V	1	1.5979	4.3803	95.6197	1.5279	15279
Playa+V	2	28.1157	1.2728	98.7272	27.7579	277579
Total		29.7137	1.4399	98.5601	29.2859	292859

Note: BC donates biogenic crust and V vegetation.

Table 6.4 Vegetation cover on sand plots of the Egyptian subset

Plot	V pixels	Vegetation %	Plot	V pixels	Vegetation %
1	18108	3.6511	4	9800	4.5533
2	11680	4.7619	5	7119	1.8681
3	6700	2.4678	Total	53407	3.1533

As can be seen in Figure 6.2, not only vegetation density, but also biogenic crust area on Egyptian subset decreases rapidly with the distance from the border. Therefore, one can logically estimate that the vegetation rate and biogenic crust percentage on the Egyptian side would be less than the average one from the measurement of the subset. In order to have a relative accurate estimation of surface composition structure on the Egyptian side, the vanishing rate of vegetation cover and biogenic crust have to be determined. Three strips with different distance from the border on the subset have been sampled for the measurement. The result is shown in Figure 6.3.

Figure 6.3 indicates that surface composition structure changes dramatically with the distance from the border. The cover age rate of sand and biogenic crust changes oppositely with the distance. At the place close to the border, their cover age rate is very similar: 43.59% for biogenic crust and 46.45% for sand. However, at the place about 800m from the border, the rate of sand increase to about 72.45% and that of biogenic crust drops to 18.03%. Therefore, one can guess that the average sand cover rate on the Egyptian side would reach up to above 80% and the average biogenic crust cover rate would decrease to below 15%. Vegetation rate also decreases, but the playa rate increase slowly with the distance. The vegetation cover rate on the strip close to the border is 11.44%, middle 7.64% and far 5.81%. The playa rate increases slowly from 2.53% in the strip close to the border through 3.23% in the middle to 3.72% in the strip far from the border. Thus, I can estimate the rate to be about 4.5% of vegetation and 3.5% for playa.

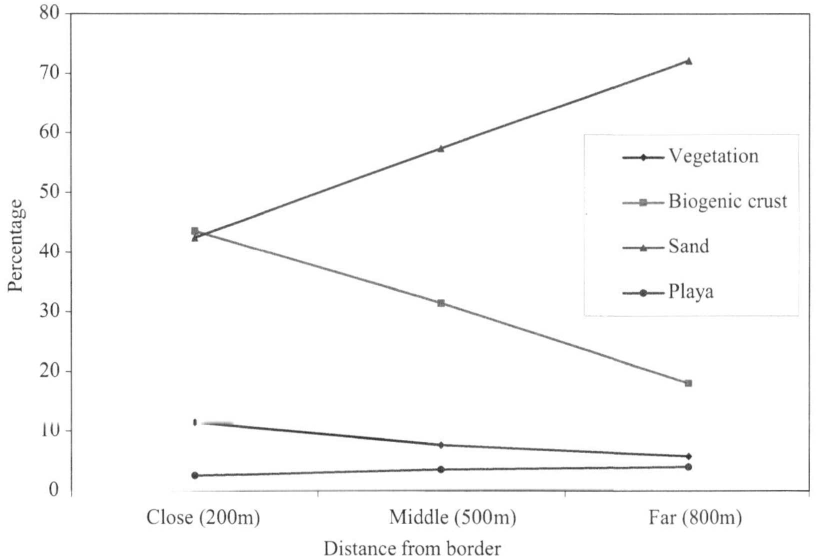

Figure 6.3 Change of surface composition on the Egyptian subset.

6.2.2 Measuring on the Israeli subset

The surface on the Israeli subset is dominated by biogenic crust mixed with shrubs, only a small part is covered with sand and playa. Thus, for simplicity, I firstly to measure the area of sand and playa plots and then use the rest as biogenic crust. Five sand dunes were sampled for the measurement. The results are listed in Table 6.5. Therefore, sand accounts for 8.89% of the total. Eight playa plots were sampled for the measurement. There is almost no vegetation on plot 1, 2, 3 and 7. Vegetation rate on group 4, 5 and 8 is very low, ranging from 1.72% in plot 8 to 2.51% on plot 5. Total pixel of the sampled playa plots is 1664364, of which vegetation occupies 2.03%. Therefore, playa accounts for 3.51% of the total on the Israeli subset. Provided vegetation accounting 17.02%, biogenic crust on the Israeli subset can be consequently estimated as 70.58%.

Table 6.5 Measurement results on the five sand dunes of the Israeli subset

Plot	Total pixel	Sand %	Vegetation %
1	551090	97.0076	3.9924
2	821591	97.2558	3.7442
3	847996	86.8448	13.1552
4	670896	90.0300	9.9700
5	595025	93.8631	6.1369
Total	3489307	92.2497	7.7503

Several plots of biogenic crust were also sampled for comparison of their difference in vegetation cover rate. The result indicates that average vegetation cover rate on interdune is 18.911% and on the dune slope 27.528%. Thus, it can be concluded that there is an obvious difference of vegetation cover rate in different biogenic crust patterns on the Israeli side. The rate in dune slope is higher than in interdune. However, the difference between dune slope bottom and middle is not very obviously. On average, the rate is about 26.60% in the bottom and about 28.88% in the middle. The higher vegetation rate on dune slope may be ascribed to the water supply from dune body. Usually the water in dune body may also be penetrated in horizontal direction, which probably makes water supply in dune slope more than in interdune where water supply only comes from vertical direction. The other reason is that the crust in the interdune, which is thicker and harder, prohibits the development of plant communities (Tielbörger, 1997).

In order to compare the spatial distribution of vegetation in different locations of the Israeli side, three sampling strips with different distance from the border were taken for the measurement of vegetation cover rate. The result shows that the rate on the strip close to the border is 16.59%, middle 18.80% and far 20.65%. This indicates that vegetation density increases with the distance from the border. Actually, due to the conservation policy, the sand dune region of the Israeli side remains in its natural evolution and is rarely disturbed by anthropogenic activities. Vegetation grows rampantly in the inner area of the region. However, the rate is different in different surface patterns. On sand and playa, the rate is only 6.87% and 2.03% respectively but on biogenic crust it is about 21.64%.

Field investigation reveals that sand on top dunes vanishes in somewhere from the border. Furthermore, the dunes with sand on top mainly concentrate in southern part of the region. The dunes in northern part rarely have continuous sand on top. They are covered with biogenic crust and sand appears as small plot. These imply that the sand cover on the Israeli side should be lower than the measurement on the subset. Considered these effects, average sand cover rate on the Israeli side is estimated to be about 7%, biogenic crust about 72% and playa about 3.5%.

6.3 Measuring vegetation cover on Landsat TM and SPOT images

Landsat TM and SPOT imagery has been extensively used for various studies of land-cover mapping (Cihlar, 2000) due to its high spatial resolution (pixel size is 30 m for TM and 10 m for SPOT under the nadir). In this study, we had six TM images and 20 SPOT images for vegetation-cover mapping in the arid region. These images were acquired over various periods and are therefore very suitable for comparing seasonal vegetation changes on each side. The TM images were acquired mainly in the mid-1990s and the SPOT images are more recent (2001-2003). Figure 6.4 is a TM image acquired on Mar 29, 1995, illustrating spatial differences in the land-cover structure in the region. The inset indicates the geographical location of the SPOT images: the area around Nizzana Research Site.

6.3.1 Method for remote sensing of vegetation cover rate

Vegetation mapping from the remote-sensing images is theoretically based on the spectral characteristics of green vegetation. The normalized difference of vegetation index (NDVI) is widely used for vegetation mapping. The index is computed from TM and SPOT images using the following formula:

$$NDVI=(IR-R)/(IR+R) \tag{6.1}$$

where IR and R respectively represents the red and infrared bands (bands 4 and 3, respectively, for TM, and bands 2 and 1 for SPOT). The higher *NDVI* value, the more vegetation on the ground. However, this index does not directly indicate vegetation cover rate. Actually, because the pixel of Landsat TM image is larger than single canopy of any plants in the arid region, *NDVI* on the pixel scale unavoidably involved the signal of both vegetation canopy and its background surface. In the sand dune region, the background is mainly biogenic crust on the Israeli side and sand on the Egyptian side.

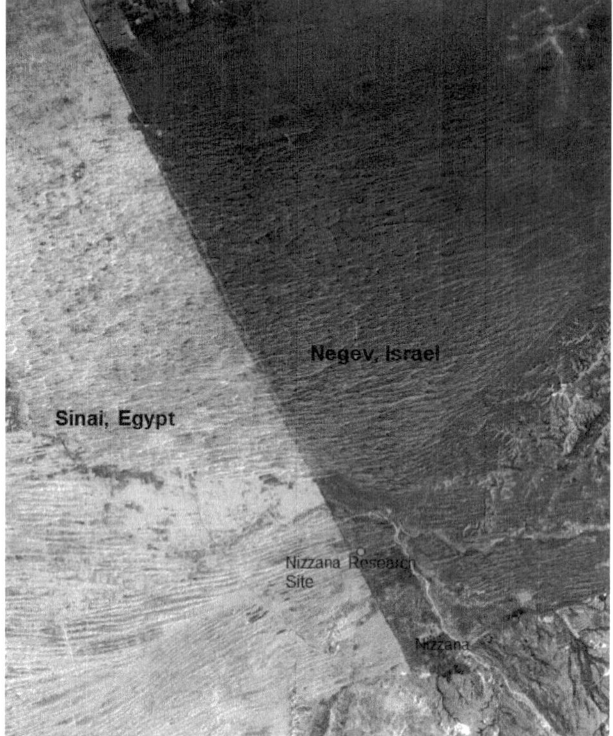

Figure 6.4 The Landsat TM image of the study region, composed of band 4, 3 and 2 as RGB. The spectral contrast is sharp on both sides of the Israel-Egypt border, which can be clearly identified with the highway along it. The sand dunes can also be seen in the image.

Because vegetation has a very high reflectance in infrared wavelengths and a low one in the red wavelength, the higher *NDVI* value, the more vegetation cover there is on the ground. However, this index does not directly indicate vegetation-cover rate. Actually, because pixels of TM/SPOT images are larger than any single canopy of plants in the arid region, *NDVI* on the pixel scale may include signal from both the vegetation canopy and the background surface. In our case, the background may be biogenic crust, bare sand or playa. However, one can expect a maximum *NDVI* value for a full vegetation-canopy cover on the ground and a minimum *NDVI* for a pure background surface where there is no vegetation cover. *NDVI* values between the maximum and minimum then represent some fraction of the vegetation cover. According to this principle, vegetation-cover rate (VR) can be calculated from NDVI as follows (Kerr et al., 1992):

$$VR = (NDVI - NDVI_b)/(NDVI_v - NDVI_b) \tag{6.2}$$

where $NDVI_v$ is the maximum NDVI, that of a fully vegetated surface, and $NDVI_b$ is the minimum NDVI, of a pure background surface without vegetation. Thus, the accuracy of using NDVI to determine vegetation cover depends on the two important coefficients: $NDVI_v$ and $NDVI_b$.

Generally, one can expect a maximal *NDVI* for the full vegetation canopy cover and a minimal *NDVI* for the pure background where there is no vegetation. *NDVI* between the maximum and the minimum represents some fraction of the surface having been covered with vegetation. According to this principle, vegetation cover rate R_v can be calculated using equation (2.21). Thus, the accuracy of using NDVI to determining vegetation cover rate will depend on the two important coefficients $NDVI_v$ and $NDVI_s$. Reflectance of green vegetation is very low in red and very high in infrared (Barett and Curtis 1978, Cracknell and Hayes 1993). According to the measurement of Tsoar and Karnieli (1996) to the spectra of desert shrub (Artemisia) in the region, the parameter $NDVI_v$ can be estimated to be about 0.60. The surface of the region is mainly covered with sand and biogenic crust (mainly cyanobacteria crust). According to the reflectance of cyanobateria crust and sand in Karnieli (1997), *NDVI* is computed as -0.04615 for sand and -0.05455 for cyanobacteria crust. On the Israeli side, the surface is mainly covered with the biogenic crust while on the Egyptian side sand is the main surface pattern. Accordingly, the *NDVI* of cyanobacteria crust and sand is used as $NDVI_s$ respectively on the Israeli side and on the Egyptian side.

Before using the above method to calculate vegetation cover rate, atmospheric correction has to be done to the TM image according to solar zenith angle, aerosol optical thickness, water vapor content and the ozone content. The following formula is generally used for the correction:

$$L_c=(aL_o-b)/(1+c(aL_o-b)) \tag{6.3}$$

where L_o is the observed radiance of Landsat TM data; L_c is the calibrated radiance; a, b and c are correction coefficients computed by the atmospheric simulation model 6S according to the factors considered. For TM3 of the image, the coefficients are a=0.00304, b=0.04626 and c=0.1102, and for TM4, a=0.00429, b=0.03019 and c=0.09093.

Then, the equation (6.1) is applied to the calibrated image for *NDVI* estimation. After generating the *NDVI* image of the region, two subsets are separately made for the Israeli side and the Egyptian side. Consequently, equation (6.2) is used to directly compute the vegetation cover rate of the two subsets. Vegetation cover rates are measured as the average value of the two subsets. According to the measurement, the vegetation rate is 5.15% on the Egyptian subset and 17.54% on the Israeli subset. This measurement result confirms that the Israeli side has much more vegetation cover. The difference of vegetation cover on both sides is up to above 12%.

6.3.2 Vegetation cover rate from the Landsat TM and SPOT images

In order to compare the spatial variation of the rate, three sampling strips paralleling to the border on both sides are sampled for the measurement: close to the border (L1), middle to the border (L2) and far from the border (L3). On each strip several boxes were sampled for calculating the mean vegetation cover rate of the location. The box is about 15×15 pixels, representing about 0.2km^2 on the ground. The results shown in Figure 6.5 clearly demonstrate the spatial variation of vegetation cover rate on both sides of the region. Two trends of vegetation distribution can be found in the region. Vegetation coverage rate on the Israeli side increases with the distance from the border (Figure 6.5a) while the rate on the Egyptian side decreases (Figure 6.5b). On the Israeli side, L1 has average vegetation rate of 15.72%, about 4% less than that of L3 (19.97%). L2 is between two extremes. However, the situation on the Egyptian side is totally opposite. L1 has the highest vegetation cover rate and L3 the lowest. This result supports the conclusion from above measurement on aerial photograph.

Another trend shown in Figure 6.5 is that the rate decreases along the southeast direction of the border. The sampling locations 1 to 6 represent a direction from northwest to southeast along the border. On both sides, it is very obvious that the rate decreases from location 1 to 6, representing the direction from northwest to southeast.

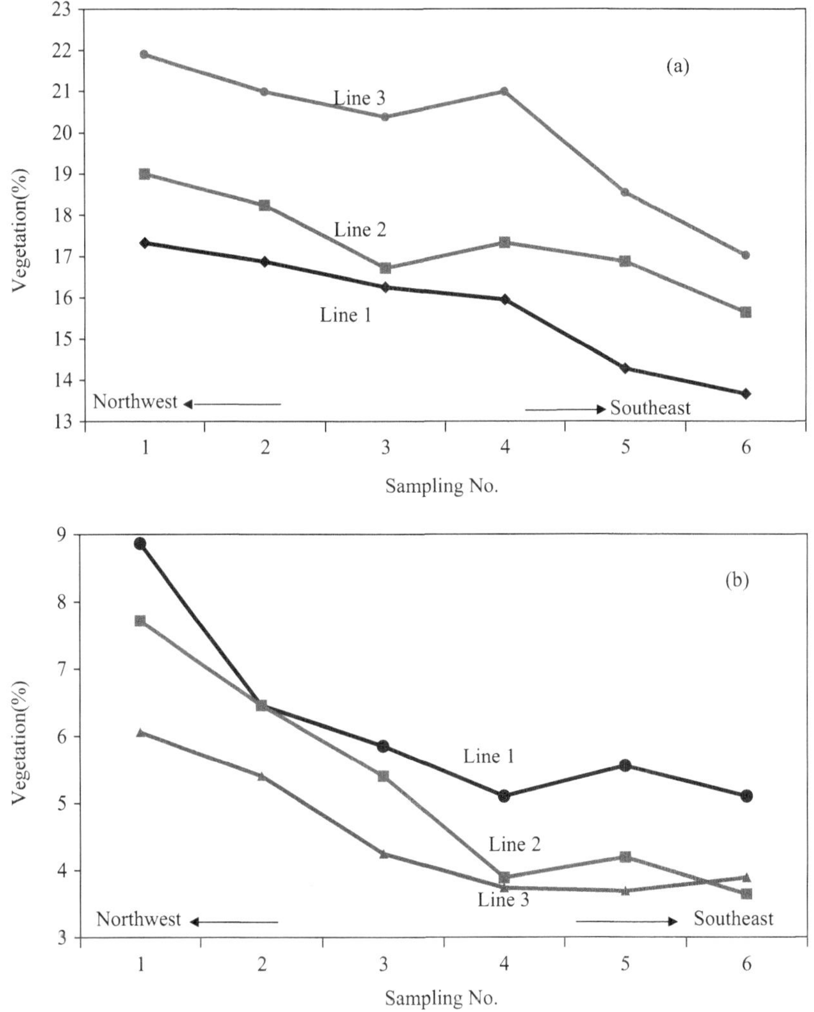

Figure 6.5 Change of vegetation cover rate with distance from the border on the Israeli side (a) and the Egyptian side (b).

Actually, this direction also indicates the distance from the Mediterranean Sea.

For detailed change of the vegetation cover rate on both sides, we took several cross-sections for the images. Figure 6.6 represents one of the results from the analysis of the TM and SPOT images. Three cross sections paralleling the border were taken for comparison of vegetation-cover changes on both sides. The changes shown in Figure 6.6 were computed from the TM image acquired on Mar 29, 1995. Since March is the blooming season for desert plants in arid regions, the changes in Figure 6.6 represent typical differences in vegetation cover across the border. The Israeli side is clearly seen to have a higher vegetation-cover rate than the Egyptian side. The cross sections were taken along the northwest to southeast direction, one close to the border, two in the middle and three further away. To reveal the general trend of vegetation-cover changes along the cross sections, each was averaged from four nearby samplings. Despite the big vibration, four features can be identified in Figure 6.6. First, vegetation rate in the northwest was generally higher than that in the southeast. For the Israeli cross sections, the rate usually changed from 22% to 30% in the northwest to 12% to 20% in the southeast. For the Egyptian cross sections, it changed from 5% to 12% in the northwest to 0 to 5% in the southeast. This can probably be attributed to distance from the Mediterranean Sea. The arid region depends mainly on the Mediterranean climate for rainfall. Within a short distance from the sea, sharp differences in rainfall magnitude can be found. Uneven distribution of rainfall in the region may shape the impacts on spatial changes in vegetation-cover density.

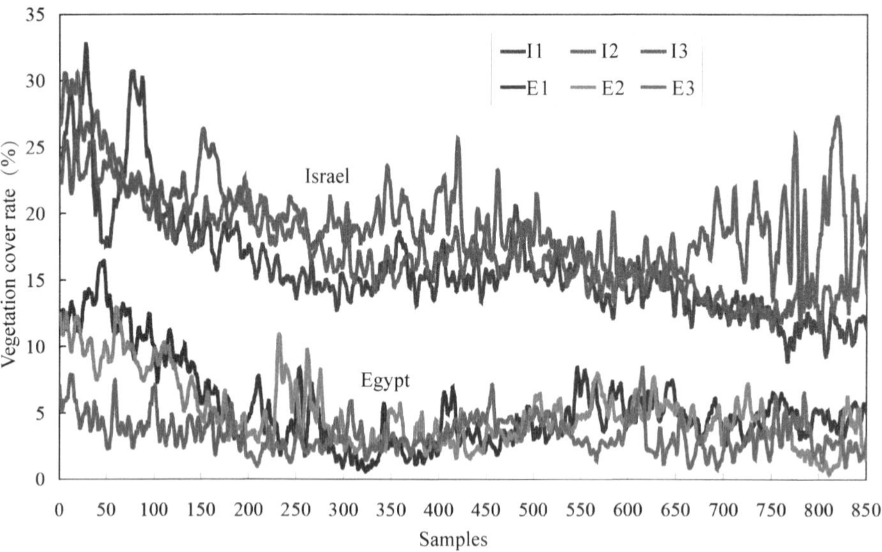

Figure 6.6 Vegetation cover changes in cross-sections of the Israeli and the Egyptian sides. I1, I2 and I3 are cross-sections 1, 2, and3 on the Israeli side; E1, E2 and E3 are cross-sections 1, 2, and 3 on the Egyptian sides. Cross-section 1 is close to the border, 2 in the middle, and 3 furthest from the border.

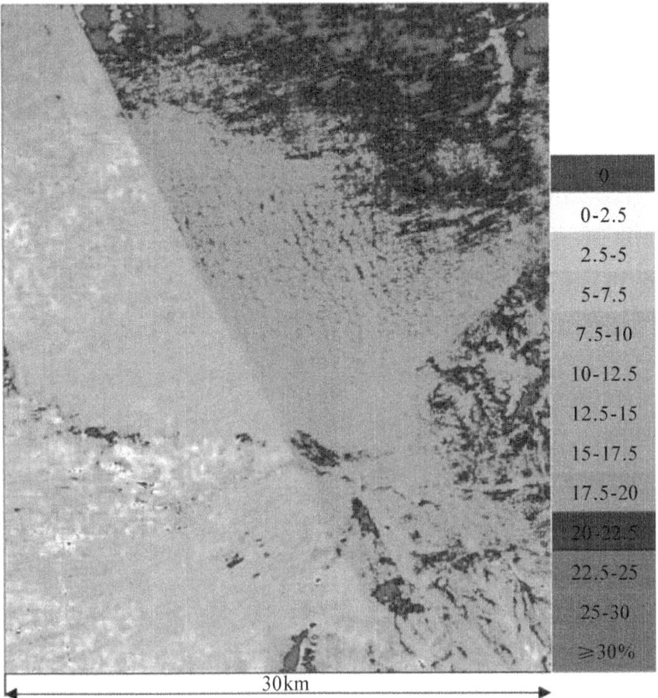

Figure 6.7 Vegetation cover rate of the Israel–Egypt border region, retrieved from Land TM image of March 29, 1995.

Second, vegetation-cover rates in the cross sections differed with distance from the border. On the Israeli side, the rates increased from 16.1% for cross section I1 (close to the border) through 17.2% for I2 (in middle) to 19.5% for I3 (furthest from the border). Opposite changes were seen on the Egyptian side, changing from 5.4% for E1 (close to the border) through 5.0% for E2 (in the middle) to 3.3% for E3 (furthest from the border). Third, the Israeli side had a much higher average vegetation cover than the Egyptian side. The average for cross sections I1, I2 and I3 was 17.6%, while that of E1, E2 and E3 was only 4.6%. This may represent the actual difference in vegetation cover across the border. Finally, spatial distribution of vegetation cover was far from even on both sides. Great vibration was a feature common to cross sections from both sides, particularly I3 and E3.

Figure 6.7 highlights the details of the spatial variation in vegetation cover in the region. In addition to the obvious difference across the border, the upper part of the

Israeli side has a higher vegetation cover than the lower part. Vegetation-cover rate is generally up to 20% to 30% in the upper part, whereas it is only 12% to 20% in the lower part. The wadi near Nizzana Research Site also has remarkably dense vegetation covering its bed. In the lower right are the small hills, where relatively higher vegetation cover is also seen in the valleys, due to more available water.

Table 6.5 and Figure 6.8 compare vegetation-cover changes in various seasons, computed from the six available Landsat TM images of the region. Seasonal changes in vegetation cover in arid regions depend strongly on rainfall. The climate in the region follows a typical Mediterranean pattern, with the rainy season generally starting in November and ending in April, despite large annual vibrations. Therefore, the period from January to March is usually the best growing season for green plants in the region. This is why we obtained much higher average vegetation-cover rates in TM images 19870404, 19950124 and 19950329. During the dry season (generally from May to October), the desert plants gradually go dormant, due to a lack of water. The plant canopy is reduced to a minimum in order to minimize water loss through evapotranspiration. Table 5 indicates that seasonal changes in vegetation cover on the Israeli side may be from above 18% in the growing season to -5% in the dry season. On the Egyptian side, the rate may be -5% in the growing season, while it may drop to below 2% in dry season.

Table 6.6 Changes of average vegetation cover rates on both sides of the arid region.

Imaging dates	No.	Israel (%)	Egypt (%)	Imaging dates	No.	Israel (%)	Egypt (%)
Landsat TM images				SPOT images			
19870404	1	16.5	4.2	20020116	7	8.7	3.9
19941121	2	7.5	1.7	20020131	8	10.4	4.4
19950124	3	16.3	4.5	20020304	9	16.9	4.9
19950329	4	18.7	4.8	20020416	10	9.0	3.3
19950617	5	10.3	3.6	20020425	11	7.2	2.6
19950921	6	8.3	3.2	20020521	12	6.2	2.9
				20021013	13	4.0	2.6
SPOT images				20021115	14	8.6	2.9
20010429	1	6.9	3.5	20021206	15	9.7	3.5
20011123	2	4.8	2.7	20021215	16	10.5	3.8

(续表)

Imaging dates	No.	Israel (%)	Egypt (%)	Imaging dates	No.	Israel (%)	Egypt (%)
20011125	3	4.1	2.5	20021230	17	13.8	4.3
20011202	4	5.9	3.1	20030110	18	14.4	4.6
20011216	5	6.1	3.4	20030201	19	16.9	4.9
20011227	6	6.6	3.5	20030227	20	18.2	5.3

Note: Imaging date is in the format of yyyymmdd. For example, imaging date 20030227 refers to February 27,2003.

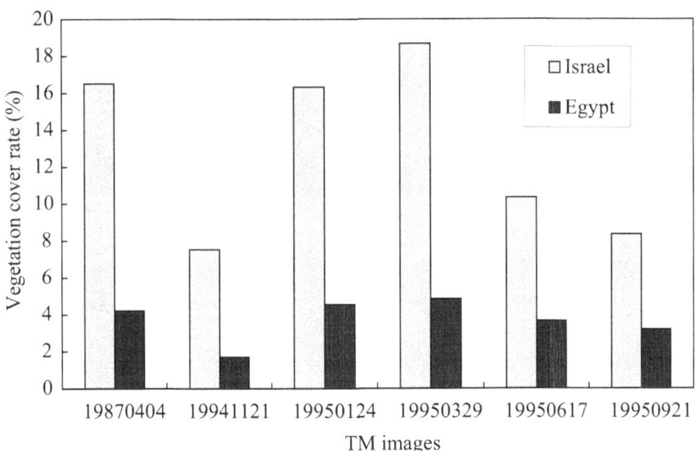

Figure 6.8 Vegetation cover rate of the Israel–Egypt border region, retrieved from Land TM image of March 29, 1995.

Seasonal changes in vegetation cover in the region are more clearly seen in the lower part of Table 5, which is computed from 20 SPOT images of the region around the Nizzana Research Site. A comparison of vegetation-cover rates on both sides is presented in Figure 6.9 and 6.10. Spatial variation is clearly seen across the border on the SPOT image (Figure 6.9) acquired on February 1, 2003. Since the imaging date was during the growing season, the vegetation-cover rate shown in Figure 6.9 reflects the region's maximal status. The rate on the Israeli side is high, up to between 22% and 30% in the dunes near the wadi. In contrast, the Egyptian side generally has a rate of below 7% (Figure 6.10) and dense vegetation can only be seen along the wadi bed, indicating that the Egyptian side is dominated by bare sand surface. Figure 8 highlights the seasonal changes in vegetation cover, where they are seen to be similar to those from the Landsat TM images. The highest vegetation-cover rates are seen in the SPOT images 9, 19 and 20, acquired during the growing season.

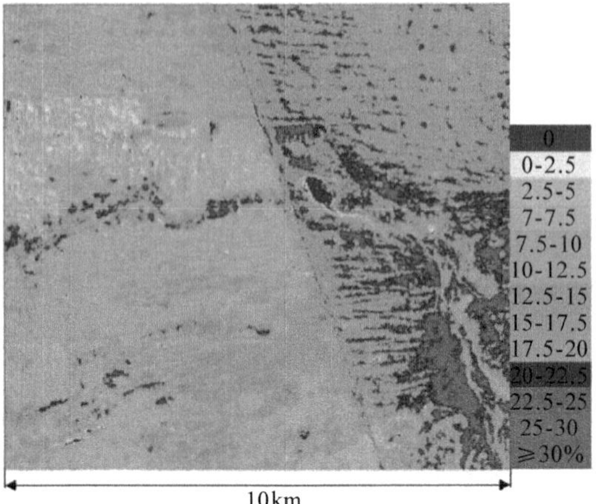

Figure 6.9 Vegetation cover rate of the Israel–Egypt border region, retrieved from SPOT image of February 1, 2003. Geographic location of the image is boxed in Figure 1.

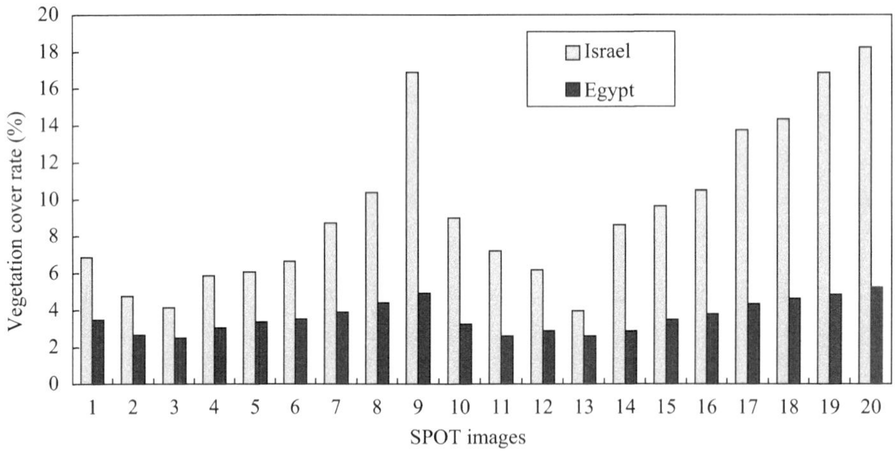

Figure 6.10 Vegetation cover changes of the arid region across the Israel–Egypt border, estimated from SPOT images. Refer to Table 5 for imaging dates.

6.4 Estimation of surface composition structure

Three methods have been employed for estimation of surface composition structure on both sides of the region. Field observation and statistics applied to the Israeli side reveals that, on average, vegetation covers about 15.63%, biogenic crust 69.42%, sand 11.94% and playa 3.01%. These results are based on the statistical analysis of the five sampling plots in the field and the calibration according to a cross-section involving two dunes and two interdunes. Total sampling area for the analysis is 3851m^2. The measurement also indicates that spatial distribution of vegetation in the region is far from homogeneity. Thus, measurement based on a larger area for estimation of surface composition structure is necessary.

An aerial photograph was used for the measurement. The photograph covers both sides of the region with an area of about 3km^2. This feature enables us to take the measurement not only on the Israeli side but also on the Egyptian side. Measurement on the aerial image indicates that a sharp contrast of vegetation cover exists on both sides. The vegetation cover rate is about 17.02% on the Israeli subset and 5.81% on the Egyptian one. The percentage of other surface patterns is also measured from the aerial image. Biogenic crust occupies about 70.58% on the Israeli subset and 18.03% on the Egyptian subset. Sand covers about 8.89% on the Israeli subset and 72.45% on the Egyptian subset. Playa only covers a small percentage on both subsets: 3.51% on the Israeli one and 3.72% on the Egyptian one.

One problem of the measurement is that the photograph only covers small area of Egyptian side (about 0.8km width) and from Figure 6.2 one can see the vegetation density and biogenic crust area decrease rapidly with the distance from the border. Three strips have been sampled for measuring the change of the cover rate of surface patterns. The results indicate that sand increases from 42.45% in strip 1% to 71.35% in strip 3 while biogenic crust drops from 43.59% in strip 1% to 18.03% in strip 3. Vegetation cover rate also drops from 11.44% in strip 1% to 5.91% in strip 3. Because strip 1 is about 200m from the border and strip 3 about 800m, one can logically expect a less biogenic crust and vegetation and a higher sand cover on the Egyptian side.

The measurement based on Landsat TM and SPOT images strongly supports the results from the aerial photograph. Because green plants have high reflection on spectral infrared range and strong absorption in the red one, *NDVI* is developed for study of vegetation in remote sensing. Using the characteristics of *NDVI* on full vegetation surface and bare soil background, vegetation cover rate of the surface can be retrieved from Landsat TM image. Measurement based on this principle indicates that vegetation cover on the Israeli subset is much denser than on the Egyptian subset. The rate is 17.54% on the Israeli side and 5.15% on the Egyptian side. Time series analysis to the SPOT images indicated that, generally speaking, average vegetation cover rate on the Israeli side can reach over 18% in the growing season, whereas it can drop to below 5% in the dry months. On the Egyptian side, the rate may increase to 5% in the growing season and drop to below 2% in the dry months. Since the surface of the Israeli side is mainly biogenic crust under canopy, and that of the Egyptian side is mainly bare sand, it is expected that the cover rate of biogenic crust on the Israeli side will change from 70% in the wet season to as high as 84% in the dry months. The rate of bare sand on the Egyptian side may range from 80% in the wet season to 82% in the dry months. Moreover, distance from border also means different vegetation cover rate on both sides. On the Egyptian side, vegetation cover rate decrease with distance from the border while the situation on the Israeli side is totally opposite. Another spatial feature of vegetation cover is that the rate decreases along the direction from northwest to southeast. This is actually the distance from Mediterranean Sea, which provides atmospheric moisture for the region.

Therefore, on the basis of the above three methods of measurement, it is reasonable to estimate the land cover structure of the region as follows. During the growing season, average vegetation cover rate may rise to above18% on the Israeli side and over 5% on the Egyptian side. However, the rate in the dry season may drop to below 5% on the Israeli side and 2% on the Egyptian side. This indicates a very high vibration in seasonal changes of vegetation cover in the region. Accordingly, biogenic crust occupies 71%-84% of the surface on the Israeli side from the wet to dry seasons, and 12% on the Egyptian side. Bare sand covers 7.5% on the Israeli side and 79%-82% on the Egyptian side. Playa accounts for 3.5% on both sides. This estimate of land cover structure has successfully used to model and explain the land surface temperature difference across the border (Qin et al., 2002b, 2005).

6.5 Modeling LST changes on both sides of the region

Because the ground surface is a mixture of various patterns, the LST retrieved from remote sensing images cannot be directly correlated into what have gained about KST change through ground truth measurement on different surface patterns. In order to understanding the relationship between ground truth measurements and remote sensing observation, consideration of ground composition structure is greatly necessary.

The combination of the four surface patterns has sharp difference on both sides. This difference has been assumed to be the direct reason leading to the sharp contrast of surface temperature obtained by remote sensing. Actually, the LST retrieved from remote sensing images is a mixed surface temperature, which can be divided into several components according to the surface composition. To the study region, total thermal radiance (I_s) emitted from the ground for remote sensing of LST can be written as:

$$I_g=A_cI_c+A_sI_s+A_vI_v+A_pI_p \tag{6.3}$$

where A_c, A_s, A_v and A_p are surface fractions of biogenic crust, sand, playa and vegetation respectively, and I_c, I_s, I_v and I_p are the thermal radiance emitted from the four surface patterns. Application of Stefan- Boltzmann law about radiance to the above equation gives

$$\varepsilon\sigma T_s^4=A_c\varepsilon_c\sigma T_{kc}^4+A_s\varepsilon_s\sigma T_{ks}^4+A_v\varepsilon_v\sigma T_{kv}^4+A_p\varepsilon_p\sigma T_{kp}^4 \tag{6.4}$$

where σ is Stefan-Boltzmann constant, T_s is the average LST, T_{kc}, T_{ks}, T_{kv} and T_{kp} are KST of biogenic crust, sand, vegetation and playa respectively. A_c, A_s, A_v and A_p are the surface fraction of the four patterns (Actually, the sum of the surface fractions is equal to 1, *i.e.* $A_c+A_s+A_v+A_p=1$), ε is the average ground emissivity, which can be simply calculated as

$$\varepsilon=A_c\varepsilon_c+A_s\varepsilon_s+A_v\varepsilon_v+A_p\varepsilon_p \tag{6.5}$$

Elimination of σ from equation (6.4) leads to the following formula for computing the average LST:

$$T_s=(F_cT_{kc}^4+F_sT_{ks}^4+F_vT_{kv}^4+F_pT_{kp}^4)^{1/4} \tag{6.6}$$

where the coefficients F_c, F_s, F_v, and F_p are the emissivity fractions, given as $F_c=A_c\varepsilon_c/\varepsilon$, $F_s=A_s\varepsilon_s/\varepsilon$, $F_v=A_v\varepsilon_v/\varepsilon$ and $F_p=A_p\varepsilon_p/\varepsilon$.

Therefore, knowing the fraction of each surface patterns and the KST change of the patterns, one can evaluate the average LST change on both sides and compare with what was obtained from remote sensing images. The evaluation requires the average emissivity on both sides. According to equation (6.5), the average emissivity can be estimated as 0.968 for the Israeli side and 0.954 for the Egyptian side. Considered the fact that the Egyptian side has much more sand, the estimation of its lower emissivity is reasonable.

During the summer, most vegetation (shrubs) in the region is in dormancy and the size of vegetation canopy reduces to its minimum. Field observation reveals that only shrub canopy can only cover about 1/3 of its ground shadow. This implies that radiance from the shrub canopy shadow is a mixed one combining with vegetation emittance and ground emittance. Considered this impact, the emissivity fraction for each surface patterns is estimated as F_c=0.81167, F_s=0.08342, F_v=0.07014 and F_p=0.034896 on the Israeli side and F_c=0.13930, F_s=0.80660, F_v=0.01830 and F_p=0.03541 on the Egyptian side. Based on the relationship, the impact of surface composition structure on LST change can be analyzed for comparison of their difference on both sides.

6.6 Impact of surface composition and ground emissivity on LST change

Because the surface is composed of four patterns and each pattern has its specific ground emissivity, the true surface temperature on both sides of the region is in fact impacted by the combination of surface composition and ground emissivity. Figure 6.11a shows the estimated combination impact of these two factors on average LST difference on both sides of the region. Two features can be clearly seen from the result shown in Figure 6.11a. First, the average LST difference on both sides decreases rapidly with the increase of temperature level. Second, the Israeli side has an average LST constantly higher than the Egyptian side only under the case that biogenic crust has an LST of above 1°C higher than sand. In summer, surface temperature of the region is usually in the range of 40-57°C at about noon during daytime. As indicated in previous chapter from ground truth measurements, biogenic crust is usually above 3°C hotter than sand. For this condition, the average LST difference on both sides will be about 2-3°C. In other words the Israeli side has an average LST of about 2-3°C higher than the Egyptian side in hot dry season. This is accordance with what was obtained on remote sensing analysis of LST change retrieved from AVHRR and Landsat TM data, as indicated in chapter 3.

In winter and during nighttime, surface temperature of the region is usually low. The difference of surface temperature between biogenic crust and sand is not as obvious as under hot dry condition. If the difference vibrates within ±1°C, one can not say that average LST on the Israeli side is higher or lower than the Egyptian side. As shown in Figure 6.11a, the average LST difference on both sides is within ±1°C for low temperature level and insignificant difference between crust and sand. This implies that in low temperature level such as several days after the heavy rain and winter nighttime the Egyptian side probably has slightly higher LST than the Israeli side. This conclusion also validates the result of remote sensing analysis to the LST change on both sides from AVHRR and Landsat TM data.

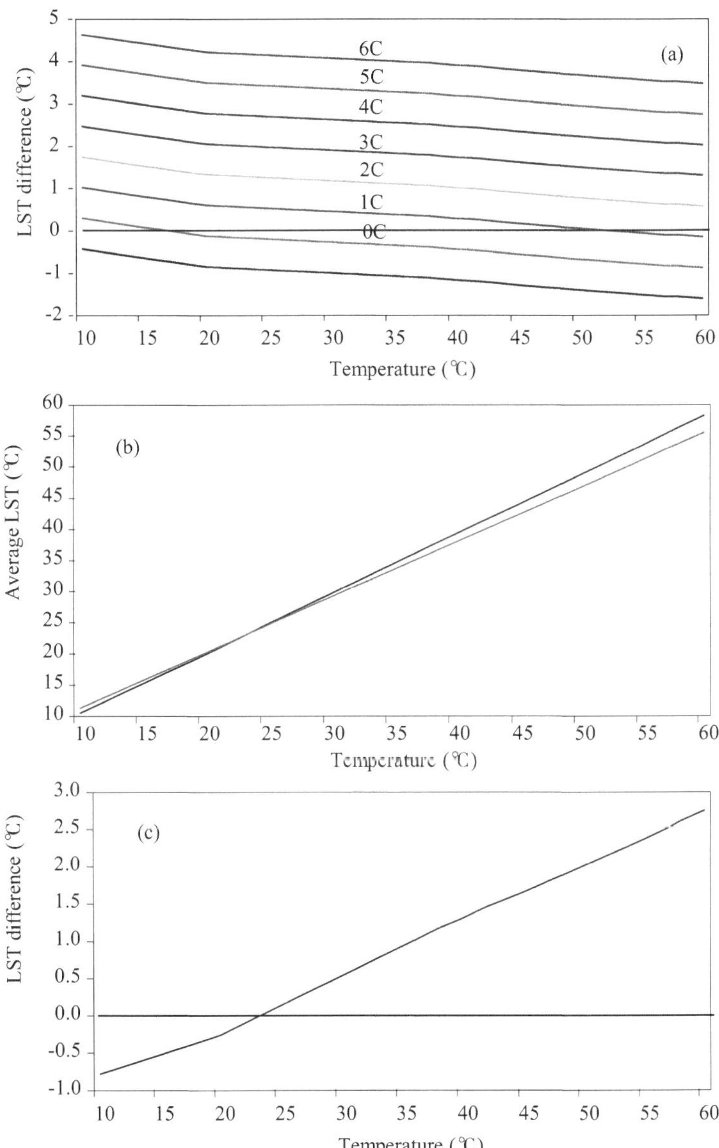

Figure 6.11 Impact of surface composition and ground emissivity on LST difference on both sides, illustrating (a) the effect of KST difference (the number in the graph) between biogenic crust and sand, (b) the possible average LST change and (c) its difference on both sides of the region.

6.7 Possible average LST change and its difference on both sides

On the basis of the above analysis, the average LST change and its difference on both sides of the region are simulated with equation (6.6) under possible temperature changes of the main surface patterns. The result is shown in Figure 6.11b and 5.6c.

As indicated by Figure 6.11b, average LST on both sides of the region changes from about 10°C to 59°C in correspondence with the temperature 10-60°C of biogenic crust. Linear characteristic of the change is eminent. Generally, the Israeli side will have higher average LST than the Egyptian side in high temperature level such as >35°C. In low temperature level such as <15°C, the Egyptian side will have steadily higher average LST than the Israeli side though the difference is still weak (<-1.0°C). Another importance feature shown in Figure 6.6b is that average LST difference on both sides is not obvious, vibrating between -0.5°C and 0.5°C, in temperature range 17-27°C. Usually LST of the region is within this range at about midnight. This is why we can not observe an obvious LST anomaly of the region on the nighttime remote sensing images of AVHRR, as indicated in chapter 3. The change of possible average LST difference between the two sides is clearly shown in Figure 6.11c. Specifically, the LST difference is high up to 2-3°C when temperature level is above 50°C, which is the general case in hot dry season

6.8 Average LST change and its difference on both sides of the region

Based on the ground truth measurements shown in Figure 4.2-4.6, the average LST change on both sides and their difference are estimated, which leads to the result given in Figure 5.7. According to Figure 4.2a, the average LST is up to 52.8°C on the Israeli side and 51.2°C on the Egyptian side, with a difference of 1.6°C. The average LST on both sides is also up to 1.6°C according to Figure 4.2b. Because the measurement shown in Figure 4.2c was taken in late afternoon, the lower average LST difference on both sides (1.1°C) is reasonable. As indicated in previous chapter, the ground truth measurements were taken at Nizzana Research Site, which is adjacent to the political border. Thus, this estimation result represents the average LST change in the area adjacent to the border, where a moderate level of average LST difference is found on both sides as shown in Figure 3.19 and 3.20.

The similar feature of LST difference on both sides of the region can be seen in early and late dry seasons. For the cases of ground truth measurements shown in Figure 4.5 and 4.6, average LST difference is found to be high up to 1.5-1.8°C (Figure 6.12).

However, the reverse LST difference on both sides exists in wet season when temperature is low. As indicated in Figure 6.11c, the Egyptian side may have higher average LST when temperature level is below 25°C. This has been perfectly validated in the ground truth measurements shown in Figure 4.3. Due to very wet surface after heavy rain, the average LST of Figure 4.3a and 4.3b is estimated to be about 19.8-22.5°C and 22.5-22.9°C on the Egyptian side. Consequently, the LST difference on both sides is -0.8°C and -0.4°C respectively.

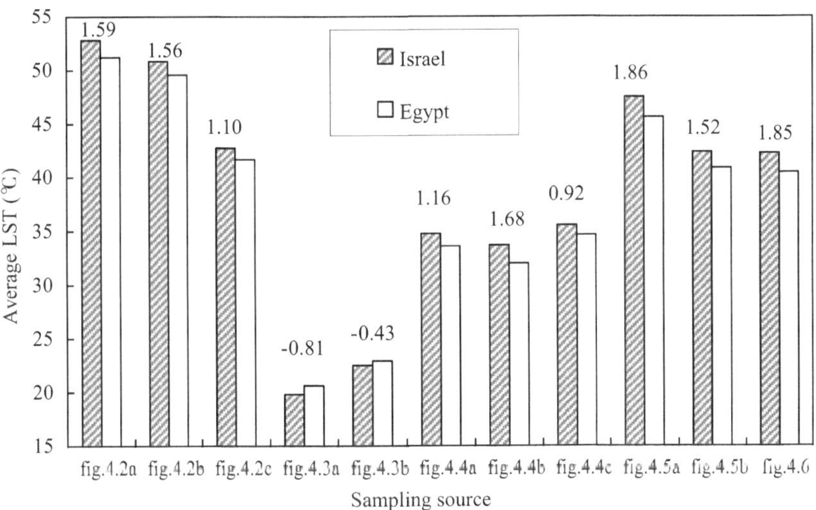

Figure 6.12 Average LST change on both sides of the region, estimated from the ground truth measurements taken at Nizzana Research Site. The number in the graph is average LST difference on both sides.

6.9 Conclusions of the analysis

The characteristics of average LST and its difference have been simulated and analyzed according to the surface composition structure on both sides of the region. First, the average LST difference on both sides is estimated for various LST differences between biogenic crust and sand. On average, it can firmly say that the Israeli side will have an obviously higher LST than the Egyptian side if biogenic crust is above 1°C hotter than sand, which occurs at moderate to high temperature level. Generally, one can expect an average LST of about 2-3°C higher on the Israeli side for high temperature level such as at noon of hot dry season. For low temperature level such as several days after heavy rain, the Egyptian side probably has a weakly higher average LST. As shown in Figure 6.6a, average LST difference on both sides decreases with temperature level.

Then, the average LST change on both sides was computed according to the possible LST change of the four typical surface patterns. The result confirms what was gained in remote sensing analysis of the LST change on both sides retrieved from AVHRR and Landsat TM data. As shown in Figure 6.11b and 5.6c, when temperature is low such as in winter, the Israeli side is weakly cooler. Obvious LST difference can be observed when temperature is at moderate to high level (>35°C). The difference tends to be minimized within the temperature range of 17-27°C, which occurs at around midnight of summer season. This explains why a sharp LST contrast on both sides can be observed in the daytime images acquired at around 14:00 (AVHRR) and 9:30 (Landsat TM) but can not be seen in the nighttime images acquired at around midnight.

Finally, the estimation method of average LST change on both sides was applied to the ground truth measurements taken on the Israeli side. As shown in Figure 6.12, the average LST on the Israeli side is obviously higher for all cases except the ones in winter when the surface was very wet and the LST was low. This confirms the above average LST estimation on both sides.

Therefore, it can be concluded that the obvious LST difference on both sides of the region is mainly due to its difference of surface composition. This surface composition difference together with their emissivity difference plays an important role in shaping LST change on both sides of the region. The Israeli side has about 72% surface covered with biogenic crust while the Egyptian side has above 4/5 surface belonging to bare sand. As indicated in previous chapter, The ground truth measurements validated that biogenic crust has much higher surface temperature than sand. The much greater fraction of biogenic crust on the Israeli side determines its much higher average LST change. And the much greater fraction of sand on the Egyptian side explains its relatively lower average LST change. Even though there is a sharp difference of vegetation cover rate on both sides, the very low fraction in estimating average LST change does not contribute a lot to even the sharp LST contrast on both sides.

7 Surface Energy balance Model for LST change

Even though the anomalous LST changes on both sides can be directly attributed to the difference of surface composition structure, the micrometeorological and hydrological mechanism governing the LST change has not been revealed. In this chapter, I want to simulate the LST change through the surface energy balance model, which has been extensively used for various research purposes (Lhomme, 1992; Dolman, 1993; Chen and Coughour, 1994; Schmugge and Humes, 1995; Dolman and Blyth, 1997; Yakirevich et al., 1997). The existing studies employed different methods to simplify the complexity of the model in terms of computation and rarely related to soil temperature change to the simultaneous soil moisture movement. In the study I intend to establelish a surface energy balance model that coupled soil temperature change simultaneously with the soil moisture movement and to develop a methodology for the numerical approximation to the model. Through this model and its numeral solution method, it is possible to estimate surface heat fluxes, surface and soil temperature change, soil water movement and other important micro-meteorological parameters that govern surface temperature change without much assumptions and simplifications to the parts of the model and its parameters. Using the meteorological data from Sede Boker in the southern Israeli desert, an example of validating the model and the methodology of its numerical solution is also presented in the chapter through comparison of the simulated soil temperature with the measured one at various depths.

7.1 Surface energy balance model

In the places where there is no heating source for the interior earth, solar radiation becomes the only source of energy controlling the most micro-meteorological events in the layer of soil-atmosphere interface. For a bare soil surface where energy storage is zero, surface energy balance at the soil-atmosphere interface can be quantitatively described as follows:

$$R_n - H - LE - G = 0 \quad (7.1)$$

where R_n is net radiation (W/m^2), H is sensible heat flux (W/m^2), LE is latent heat flux (W/m^2), G is soil heat flux (W/m^2). Because 1W=1J/s, the dimension of the terms in equation (7.1) can also be written as Js^{-1}m^{-2} so that it is compared to the unit used in the following computation. According to the equation, net radiation R_n is equal to the sum of sensible heat flux H, latent heat flux LE and the soil heat flux G. The terms in equation (7.1) are all related to surface temperature, which becomes the key factor for numerical solution of the model.

7.1.1 Net radiation

The short-wave radiation from the sun reaches the top of the atmosphere at about 1395W/m^2. As it passes through the atmosphere, the solar radiation is scattered, absorbed and reflected by different types of molecules and colloidal particles. Thus, the global short-wave radiation reaching the ground surface consists of direct solar radiation and diffuse sky radiation (Brutsaert, 1982). At the ground surface, part of the global short-wave radiation R_s is reflected by the surface back into atmosphere at a density that depends on the albedo of the surface. At the same time, the Earth's surface also emits some long-wave radiation into the atmosphere and the atmosphere also emits some long-wave radiation that reaches the ground surface. Some of the incoming atmospheric long-wave radiation is reflected by the ground surface back into the atmosphere. Therefore, at the ground surface, the net radiation can be expressed as follows after applying Stefan-Boltzmann law in the emitted terms of atmosphere and the ground:

$$R_n = R_s(1-\rho) + \varepsilon_a \sigma T_a^4 - \varepsilon_s \sigma T_s^4 - (1-\varepsilon_s)\varepsilon_a \sigma T_a^4 \quad (7.2)$$

where R_n is net radiation (W/m^2), R_s is global hemispheric radiation (W/m^2), ρ is surface albedo, T_s is surface temperature (K), T_a is air temperature (K), ε_s and ε_a are surface and air emissivities respectively, σ is Stefan-Boltzmann constant ($\sigma = 5.67 \times 10^{-8}$ Wm^{-2}K^{-4}).

Studies indicated that air emissivity ε_a is strongly coupled to air vapor pressure (e_a) and air temperature (T_a) near the surface (Brutsaert, 1982):

$$\varepsilon_a = 1.24(e_a/T_a)^{1/7} \quad (7.3)$$

Here e_a is in mb (1mb=100Pa) and T_a in K. Because air vapor pressure also has strong correlation with air temperature, ε_a also can be approximated by the empirical formula of Swinbank (1963):

$$\varepsilon_a=0.92\times10^{-5}T_a^2 \tag{7.4}$$

Both formulap (7.3) and (7.4) have been found to yield satisfactory results with daily means at mid-latitudes and at temperature above 0°C, as well described by a standard atmosphere (Brutsaert, 1982).

7.1.2 Sensible heat flux

Successful estimation of sensible heat flux is essential for the application of surface energy balance model (Lhomme et al., 1994). The amount of sensible heat flux H strongly depends on surface-air temperature difference and air resistance to heat transfer, calculated by the following formula (Verma and Barfield, 1979; Schmugge and Humes, 1995):

$$H=\rho_a c_a(T_a-T_s)/r_a \tag{7.5}$$

where ρ_a is density of air (ρ_a=1.205 kg/m^{3}at 20°C), c_a is specific heat of air (c_a=1005 J kg^{-1} K^{-1}), r_a is air resistance to heat transfer (s/m), which is given as (Berliner, 1988)

$$r_a=\frac{(\ln(z/z_0)-\varphi_h)}{kU} \tag{7.6}$$

$$U=\frac{w_z k}{\ln(z/z_0)-\varphi_m} \tag{7.7}$$

where u_z is wind velocity (m/s) at standard height z (z=2m), k is Von Karman constant (k=0.4), z_0 is roughness length (m), φ_h and φ_m are the stableility correction parameters for heat and momentum, which can be estimated through the ratio function of standard height to Monin-Obhukov length (Berliner, 1988; Courault et al., 1996)

$$z/L_m=\frac{-kzgH}{\rho_a c_a U^3} \tag{7.8}$$

where L_m is Monin-Obhukov length and g the acceleration of gravity (g=9.8m/s^2). The stableility correction parameters are defined as follows (Berliner, 1988; Courault et al., 1996):

If $z/L_m<0$, then

$$\varphi_h=2\ln((1+X)/2)+\ln((1+X^2)/2)-2\arctan(X)+\pi/2$$

$$\varphi_m=2\ln((1+X^2)/2)$$

If $0<z/L_m<\ln(z/z_0)$, then $\varphi_h=\varphi_m=-5z/L_m$

If $z/L_m>=\ln(z/z_0)$, then $\varphi_h=\varphi_m=-5\ln(z/z_0)$

where $X=(1-16z/L_m)^{1/4}$ and π is circle constant (π=3.14159265) Therefore, the estimation of sensible heat flux requires an iterative calculation in the simulation process. The iteration could be as follows.

(1) Let $\varphi_h=\varphi_m=0$, $z/L_m=0$ and $H_1=0$.

(2) Use equation (7.6) to calculate r_a.

(3) Use equation (7.5) to calculate H.

(4) Use equation (7.8) to calculate z/L_m.

(5) Compare H with H_1, if $H1-H$ is small enough to reach the required accuracy, then stop the iteration and get the final result. If not, let $H_1=H$ and compare z/L_m with $\ln(z/z_0)$ to determine new value of φ_h and φ_m.

(6) Repeat (2) to (5) until the required accuracy is reached.

7.1.3 Latent heat flux

The availability of energy and moisture at the earth-atmosphere interface is the critical conditions for evaporation. The energy required for evaporation is generally termed as latent heat flux LE, which can be computed by the following formula (Schmugge and Humes, 1995)

$$LE=0.622L(e_a-e_s)/(P_a(r_a+r_s)) \tag{7.9}$$

where e_a and e_s are air and surface vapor pressures (kPa) respectively. P_a is atmospheric pressure (P_a= 101.325 kPa). L is latent heat vaporization ($L=2.543\times10^6$ J/kg). r_s is surface resistance (s/m), given empirically as (Berliner, 1988)

$$r_s=100(0.413\theta_s/\theta)^{1.5} \tag{7.10}$$

where θ is volumetric soil water content (kg/m^3) and θ_s is soil water content at saturation (kg/m^3).

Actually, evaporation from ground surface must be equal to the change of soil water content in the profile. Thus, we have

$$LE-LE_c=0 \tag{7.11}$$

where LE_c donates the energy (W/m^2) caused by the change of soil water content in time interval ∂t. A complete description of soil water content change in the profile should be in three dimensions. However, in many cases, soil water movement over planar directions can be assumed to be negligible small (Brutsaert, 1982; Rose, 1969). This is especially true in many arid environments where soil water content is extremely low and the planar difference of available energy for evaporation is not significant in short distance. Under this assumption, the energy used for driving soil water movement in time interval ∂t can be computed as follows.

$$LE_c = L\int_0^z \frac{\partial\theta}{\partial t} dz \tag{7.12}$$

where t is time (s) and z is the depth (m) of soil profile under consideration. $\partial\theta/\partial t$ donates the rate of soil water change. When the planar movement is neglected, the vertical movement of soil water can be described by the following differential equation (Berliner, 1988):

$$\partial\theta/\partial t = \partial(K_c\partial\psi/\partial z)/\partial z + \partial(gK_c)/\partial z + \partial(hsD_v\partial T/\partial z)/\partial z + \partial(e_vD_v\partial h/\partial z)/\partial z \tag{7.13}$$

where K_c is soil hydraulic conductivity (kg s/m^3), h is relative humidity of the gas filled in the soil pore, ψ is soil water potential (J/kg=m^2/s^2), s is slope of saturated vapor pressure *vs.* temperature (kPa/K), e_v is saturated vapor pressure (kPa) and D_v is apparent vapor diffussivity (kg m^{-1}s^{-1}kPa^{-1}). Soil water potential is coupled with soil vapor pressure and temperature *via* equation

$$\psi = RT\ln(e/e_v) \tag{7.14}$$

In which R is universal gas constant (R=461.52 J kg^{-1} K^{-1}). Actually, soil relative humidity h is calculated as the ratio of soil vapor pressure e and saturated soil vapor pressure e_v, *i.e.*

$$h = e/e_v \tag{7.15}$$

Thus, the relationship between soil water potential and relative humidity is given as $\psi = RT\ln(h)$.

The first two terms of equation (7.13) describe the soil water movement in liquid phase due to potential difference and gravitation respectively. The last two terms describe the movement of soil water in the vapor phase due to temperature and potential gradients respectively. Therefore, the change of soil water content with time is described as the function of such important variables as soil water potential, soil relative humidity and soil temperature.

7.1.4 Soil heat flux

Soil heat flux can be computed from soil temperature change and soil heat capacity in the profile. Usually, soil heat transfer in planar directions is negligibly small and only in vertical direction is practically worthy of consideration. Thus, the term G in (7.1) can be computed as (Kimball and Jackson, 1979):

$$G = \int_0^z C_s \frac{\partial T}{\partial t} dz \tag{7.16}$$

where $\partial T/\partial t$ donates the rate of soil temperature change and C_s is volumetric soil heat capacity ($Jm^{-3}K^{-1}$) which can be expressed as $C_s=\rho_s c_s$, in which ρ_s is soil density (kg/m^3) and c_s specific heat of soil ($Jkg^{-1}K^{-1}$). However, specific heat of soil strongly depended on soil properties especially the materials constituting the soil. It is also highly variable, depending on the change of soil water content. Usually, the soil can be viewed as constituted by soil minerals (mainly sand, silt and clay), organic materials, water and air in different proportions. Generally, the specific heat of these soil constituents is stable for practical purposes. Thus, it was suggested that the heat capacity of the soil could be computed through the thermal properties of its constituents with links to their fractions (Brustaet, 1982; Marshall and Holmes, 1979).

$$C_s=\rho_w c_w V_w+\rho_q c_q V_q+\rho_m c_m V_m+\rho_o c_o V_o+\rho_a c_a V_a \tag{7.17}$$

where ρ is density of soil constituents, c is specific heat, V is volumetric fraction, the subscripts w, q, m, o, a are referred to water, quartz, clay mineral, organic materials and air respectivelyl.

Natural soil is generally viewed as a poor electrical conductor (Rose, 1969). This is especially true when soil water content is very low and the soil is loose or with large porosity. In this case, heat transfer in the soil by thermal conduction is ascribed to net molecular exchange of kinetic energy, which takes place from the more energetic molecules (hotter regions) to those cooler regions where the molecular motion is less energetic (Rose, 1969). In the other hand, heat transfer in the soil also causes the imbalance of energy distribution in the soil, which drives soil water movement especially in vapor form of unsaturated state. And soil water movement also eases the process of soil heat transfer. Thus, a complete description of soil heat flux must consider the equilibrium of these two factors. In the time interval ∂t, the intensity of soil heat transfer can be described by the following equation (Berliner, 1988):

$$C_s\partial T/\partial t=\partial(K_s\partial T/\partial z)/\partial z+\partial(hsLD_v\partial T/\partial z)/\partial z+\partial(e_v LD_v\partial h/\partial z)/\partial z \tag{7.18}$$

Where K_s is soil thermal conductivity ($W\ m^{-1}\ K^{-1}$). The first term on the right hand side of equation (7.18) describes the heat transfer due to temperature gradients. The second term is the energy due to soil water vapor movement under the temperature gradient. The third term is the energy due to the change of soil water vapor distribution under the gradient of the vapor distribution in the soil.

7.2 Soil parameters of the model

The solution of above surface energy balance model requires the estimation of soil parameters involved in (6.13) and (6.18). However, most of these parameters do not have a fixed relationship with each other, but depend on soil properties, especially the structure, components and textures (Hanks and Ashcroft, 1980). The following empirical relationships were proposed for general purposes of estimating soil parameters required for modeling (Campbell, 1985).

7.2.1 Soil thermal conductivity

Soil thermal conductivity K_s is defined as the heat flux density conducting through the soil divided by the temperature gradient (Koorevaar et al., 1983). The parameter is extremely important because it strongly impacts the speed of heat transfer in the soil, which shapes soil temperature change and soil moisture movement. Like thermal capacity, thermal conductivity of natural soil is also highly variable and depended on soil properties, especially bulk density, water content, quartz content and organic matter content (Kimball and Jackson, 1979; Campbell, 1985). This analogy provides the similar way of computing K_s from its constituents. However, soil thermal conductivity does not have a simple relationship with the thermal conductivity of individual soil constituents because the conduction of heat takes place through all kinds of sequences of the conducting materials, in series and parallel (Koorevaar et al., 1983). The value of K_s depends highly on the way in which the best conducting mineral particles are interconnected by the less conducting water phases and are separated by the poorly conducting gas phase. Thus, shape factors have to be considered for the computation of K_s (de Vries, 1963; Yakirevich et al., 1997; Berliner, 1988):

$$K_s=\frac{F_wV_wK_w+F_qV_qK_q+F_mV_mK_m+F_oV_oK_o+F_aV_aK_g}{F_wV_w+F_qV_q+F_mV_m+F_oV_o+F_aV_a} \tag{7.19}$$

where V is volumetric fraction of soil constituents, K is thermal conductivity, F is shape factors, and the subscripts w, q, m, o, a and g are referred to water, quartz, clay minerals, organic matters, are and gas respectively in the soil. The thermal conductivity of the gas filled the soil pore K_g and the shape factors can be computed as follows:

$$K_g=K_a+K_v \qquad K_v=0.075\,e/P$$

$$F_w=V_sC_w/C \qquad F_q=V_sC_q/C$$

$$F_m=V_sC_m/C \qquad F_o=V_sC_o/C$$
$$F_a=V_a$$

where K_a is air thermal conductivity, K_v is the apparent thermal conductivity of the gas filled the soil pore, K_v=0.075e/P, where e is vapor pressure and P is total gas pressure, V_s is the volumetric fraction of soil solids (V_s=1-V_a) and C_w, C_q, C_m, C_o and C are the constants, with C_w=1, C_q=0.051, C_m=0.104, C_o=1.298 and $C=C_w+C_q+C_m+C_o$=2.543.

At low water content, the air space controls the thermal conductivity of the soil and all types of soils such as litter, sand and silt loam have similar thermal conductivity. At high water content, the thermal conductivity of the solid phase becomes more important and the differences of bulk density and soil composition results in significant difference of thermal conductivity. The transition of soil thermal conductivity from low to high occurs at low water content in sand and at high water content in soils with high clay content. This relationship provides a quantitative way of computing soil thermal conductivity for modeling and the following empirical formula was proposed (Campbell, 1985; McInnes, 1981):

$$K_s=A+2.8V_s(\theta/\rho_w)^2+(A-B)\exp(-(C\theta/\rho_w)^4) \tag{7.20}$$

where ρ_w is water density ($\rho_w\approx$1000kg/m^3) and A, B and C are parameters given as:

$$A=(0.57+1.73V_q+0.93V_m)/(1-0.74V_q-0.49V_m)-2.8V_s(1-V_s)$$
$$B=0.03+0.7V_s^2 \tag{7.21}$$
$$C=1+2.6/(V_c)^{1/2}$$

where V_s is volumetric fraction of solids in soil, *i.e.* $V_s=V_q+V_m$.

The relationship between soil water potential and soil water content varies in different soil types (Rose, 1969; Campbell, 1985). However, the relationship does exist for specific soil. This can be expressed by water retention curve. Campbell (1985) suggested the following functional relationship between the potential and the content:

$$\psi=\psi_0(\theta/\theta_s)^{-b} \tag{7.22}$$

In which ψ_0 is the soil water potential at saturation. Moreshet et al. (1983) defined ψ_0=-4.7J/kg for their soil. Actually ψ_0 ranges from -9J/kg to -0.6J/kg and it is rational to give ψ_0=-5J/kg for many cases (Compbell, 1985). The parameter b is also strongly depended on soil properties, especially texture and Moreshet et al. (1983) gave it as b=10.6 for their soil and Berliner (1988) b=2.5 for his. According to Campbell (1985), b can range from 24 to 2 and be estimated by the following empirical formula:

$$b=d_g^{-1/2}+0.2\sigma_g \quad (7.23)$$

where d_g and σ_g are geometric mean particle diameter (mm) and geometric standard deviation respectively, given as

$$d_g=\exp(M_c\ln D_c+M_c\ln D_c+M_c\ln D_c)$$

$$\sigma_g=\exp\{[M_c(\ln D_c-d_g)^2+M_l(nD_l-d_g)^2+M_d(\ln D_d-d_g)^2]^{1/2}\}$$

where D_c, D_l and D_d are arithmetic mean diameter (mm) of clay, silt and sand, generally D_c<0.002mm, 0.002≤D_l<0.05mm and 0.05≤D_d<2.0mm, M_c, M_l and M_d are the mass fraction of clay, silt and sand respectively in the soil.

7.2.2 Soil hydraulic conductivity

Soil hydraulic conductivity K_c has functional relationship with soil water content (Marshall and Holmes, 1979; Braud, 1998) though the determinants of the parameter are the properties of soil but not only water content. Studies indicated that different soils have different hydraulic conductivity under the same water content (Brutsaert, 1982; Marshall and Holmes, 1979; Campbell, 1985). However, for the same soil, soil water content is the main determinant of hydraulic conductivity and the following relationship between the conductivity K_c and soil water content θ was proposed for practical calculation (Campbell, 1985; Braud, 1998):

$$K_c=K_h(\theta/\theta_s)^m \quad (7.24)$$

where K_h is saturated hydraulic conductivity and m is a shape parameter related to the texture of the soil (Choudhury and Federer, 1984). Empirically, Campbell (1985) gave m as $m=2+3/b$ and Braud (1989) $m=2+1/b$. As suggested by Campbell (1985), soil saturated hydraulic conductivity K_h is calculated as

$$K_h=0.002\exp(-4.26(M_l+M_c)) \quad (7.25)$$

where V_s and V_c are the mass fractions of silt and clay in the soil, respectively. Generally speaking, hydraulic conductivity K_c is extremely small when soil water content is low. According to the measurement of Gardner (1956), K_c of a sandy loam soil is about 5.4×10^{-12} kg s/m^3 for θ=110 kg/m^3 and 1.3×10^{-10} kg s/m^3 for θ=170 kg/m^3. According to the graph in Campbell (1985), K_c is about 2×10^{-9} kg s/m^3 for Guelph loam with water content 200kg/m^3 and about 2×10^{-5} kg s/m^3 for Botany sand with water content 100kg/m^3.

Apparent vapor diffusivity D_v is given by Troeh et al. (1982) as

$$D_v=D_a(0.622\rho_a/P)[(V_a-u)/(1-u)]^v \quad (7.26)$$

Where P is total gas pressure (kPa), D_a is vapor diffussivity (m^2/s) in the air, given by $D_a=(2.22+0.158T_c)\times10^{-5}$ in which T_c is the temperature in °C, u and v are parameters, given by Toeh et al. (1982) as $u=0.05$ and $v=1.5$.

Saturated vapor pressure e_v is a function of temperature, given as

$$e_v=[\exp(26.6904-6109.74/T-0.00916189T]/10 \tag{7.27}$$

in which T is temperature in K. And the parameter s can be calculated as

$$s=s_t e_v/(T^2) \tag{7.28}$$

where s_t is a temperature constant (s_t=5307K).

When the above soil parameters and such meteorological data as global radiation, air temperature, air relative humidity and wind speed are available, the surface energy balance model described in (6.1.1) can be numerically solved for estimation of heat flux and surface temperature.

7.3 Numerical solutions to the simulation model

Numerical solution of the model is based on the following approximations: iterative computation of latent heat flux and surface temperature respectively from equation (7.13) and (7.1), simultaneous solution of soil temperature and soil moisture change from differential equation (7.13) and (7.18).

7.3.1 Newton-Raphson method for approximation

As we have seen, the relations in the surface energy balance model are very complicated. It is impossible to directly express the key factor (surface temperature T_s) of the model as a function of other variables. Usually, to solve such complicated equation as our model involves the application of numerical approximation method. Among the approximation methods, Newton-Raphson iterative technique is a good one because it can rapidly reach the solution with the required accuracy. Figure 7.1 illustrates the principle of Newton-Raphson iterative techniques for approximation of a complicated function. For our surface energy balance model (7.1), we can take it as a function of surface temperature (T_s) because all terms of the model changed in the surface temperature. Thus we can rewrite the model as follows:

$$f(T_s)=R_n-H-LE-G=0 \tag{7.29}$$

This function is so complicated, but it should have the form as one of the two curves illustrated in Figure 7.1. This means that there is surface temperature T_s to make the $f(T_s)=0$, as shown in Figure 7.1. In order to solve for this surface temperature, we have to try several times for different surface temperature to reach this solution of surface temperature. Using the Newton-Raphson approximation technique, we can quickly reach this solution.

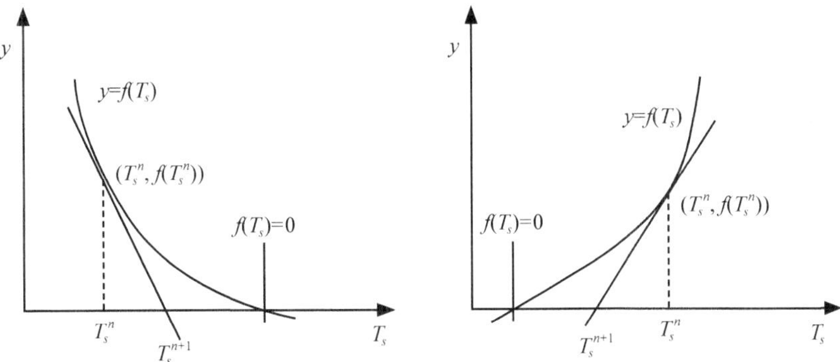

Figure 7.1 The principle of Newton-Raphson approximation to equation

Since the above equation (7.29) is a successive function within such interval as $T_s \subseteq (-50°C, 100°C)$. Therefore, to apply Newton-Raphson iterative method for approximation, we start with the point T_s^0 in the interval $T_s^0 \subseteq (-50°C, 100°C)$ for the solution of T_s and proceed to determine additional approximation by

$$T_s^{n+1} = T_s^n - \frac{f(T_s^n)}{f'(T_s^n)} \qquad (n=0,1,2,......) \tag{7.30}$$

where $f(T_s^n)$ and $f'(T_s^n)$ are the values of the function $f(T_s)$ and its derivative at T_s^n. Geometrically, Newton-Raphson method means that the tangent at the point $(T_s^n, f(T_s^n))$of the curve $y=f(T_s)$ is extended to the intersection with T_s axis at T_s^{n+1} (Figure 6.1), which is used as the new approximation to the solution T_s. If T_s^n is a good approximation, it can be expected that T_s^{n+1} approximate T_s still better (Flanders and Price, 1978) because T_s^{n+1} is much closer to T_s. Therefore, after iterative calculation many times when $|f(T_s^n)| \rightarrow 0$ or is less than the required accuracy, we get $T_s^{n+1} = T_s^n \rightarrow T_s$ and stop the iterative calculation. Thus, we can conclude that T_s^n is the solution of surface temperature T_s from the model under the balance of available inputs.

If δT is taken to be small enough, the derivative $f'(T_s^n)$ can be given as

$$f'(T_s^n) = \frac{f(T_s^n + \delta T) - f(T_s^n)}{\delta T} \tag{7.31}$$

Thus, we have

$$T_s^{n+1}=T_s^n-\frac{f(T_s^n)\delta T}{f(T_s^n+\delta T)-f(T_s^n)} \tag{7.32}$$

To solve the function $f(T_s)$ for T_s also involves to solve the differential equations (6.15) and (6.3.16) simultaneously. Crank-Nicholson technique can be used to approximate the solution of the two different equations. This technique is mathematically complicated. A detailed description will be given in section 5.

In order to compute latent heat flux LE for the approximation of surface temperature T_s, equation (7.11) has also to be solved for vapor pressure e_s or relative humidity h_s of the ground surface. And the solution of this equation is coupled to the simultaneous solution of equation (7.9) and (6.12), of which the latter is resulted from differential equation (7.13). Due to impossible to give a direct solution of surface humidity h_s for the computation, approximation has to be employed for the iterative calculation of h_s from equation (7.11). According to equation (7.9), the unknown for computing LE is surface vapor pressure e_s, which is related to surface relative humidity h_s through equation (7.16). Therefore, both LE and LE_c can be viewed as the function of h_s and we can rewrite equation (7.11) as

$$f(h_s)=LE-LE_c \tag{7.33}$$

Similarly, Newton-Raphson iterative approximation method can be used to give a solution of h_s from this equation. The procedure of the solution is the same as that used for the solution of T_s from equation (7.29) hence not necessary to repeat.

7.3.2 Derivation of the differential equations

Using Newton-Raphson approximation to solve surface temperature and latent heat flux from equation (7.1) and (7.11) involves the solution of differential equation (7.13) and (7.18). In order to solve differential equation (7.13) about soil water movement, we have to derive the equation into a proper form for approximation. According to equation (7.14) and (7.15), we can rewrite the derivative $\partial\psi/\partial z$ in equation (7.13) as

$$\begin{aligned}\partial\psi/\partial z&=\partial(RT\ln(h))/\partial z=RT\partial(\ln(h))/\partial z+R\ln(h)\partial T/\partial z\\&=(RT/h)\partial h/\partial z+R\ln(h)\partial T/\partial z\end{aligned} \tag{7.34}$$

Thus, the first differential term in the right hand side of equation (7.13) can be rewritten as

$$\partial(K_c\partial\psi/\partial z)/\partial z=\partial((RK_cT/h)\partial h/\partial z)/\partial z+\partial(RK_c\ln(h)\partial T/\partial z)/\partial z \quad (7.35)$$

At the same time, the term $\partial\theta/\partial t$ can be written as

$$\partial\theta/\partial t=(\partial\theta/\partial h)\partial h/\partial t \quad (7.36)$$

Let $C_h=(\partial\theta/\partial h)$, we have

$$\partial\theta/\partial t=C_h\partial h/\partial t \quad (7.37)$$

The acceleration of gravity is a constant. Thus, the second terms in the right side of equation (7.1.13) can be simply rewritten as

$$\partial(gK_c)/\partial z=g\partial K_c/\partial z \quad (7.38)$$

Therefore, equation (7.1.13) can be rewritten as

$$C_h\partial h/\partial t=\partial((RK_cT/h)\partial h/\partial z)/\partial z+\partial(RK_c\ln(h)\partial T/\partial z)/\partial z+g\partial K_c/\partial z+ \partial(hsD_v\partial T/\partial z)/\partial z+\partial(e_vD_v\partial h/\partial z)/\partial z \quad (7.39)$$

Reorganizing the right hand side, we get

$$C_h\partial h/\partial t=\partial((RK_c\ln(h)+hsD_v)\partial T/\partial z)/\partial z+ \partial((RK_cT/h+e_vD_v)\partial h/\partial z)/\partial z+g\partial K_c/\partial z \quad (7.40)$$

For simplification, we define

$$K_a=RK_c\ln(h)+hsD_v \quad (7.41)$$

$$K_b=RK_cT/h+e_vD_v \quad (7.42)$$

where both K_a and K_b have the same dimension as $JsK^{-1}m^{-3}$. Substituting K_a and K_b into equation (7.3.12), we get

$$C_h\partial h/\partial t=\partial(K_a\partial T/\partial z)/\partial z+\partial(K_b\partial h/\partial z)/\partial z+g\partial K_c/\partial z \quad (7.43)$$

Similarly, equation (7.1.18) can be reorganized as

$$C_s\partial T/\partial t=\partial(K_d\partial T/\partial z)/\partial z+\partial(K_e\partial h/\partial z)/\partial z \quad (7.44)$$

where K_d and K_e are defined as

$$K_d=K_s+hsLD_v \quad (7.45)$$

$$K_e=e_vLD_v \quad (7.46)$$

The dimension of K_d and K_e is $Jm^{-1}s^{-1}$. With initial and boundary conditions, all the variables in (7.43) and (7.44) are known for time interval j (j=0,1,2,3...) and what we need to solve from the two equations is that the value of the variables for the time interval j+1. Therefore, initial and boundary conditions are critical for the numerical solution of the two equations.

7.3.3 Initial and boundary conditions

Equation (7.43) and (7.44) cannot be solved except the initial and the boundary conditions are given. For the time interval $j+1$, we can use the value of time interval j for initial conditions. Thus, at the beginning, an initial condition is generally assumed for the solution. And after the value of $j+1$ is computed, we use it as the initial condition for the next iteration of computation until the time we expect to stop.

There are several ways to give an initial condition at the beginning of computation. Usually it is determined according to the measurement of soil temperature and soil water content at different depths of the profile under consideration.

It is the fact that soil temperature and soil water content remain constant at specific depth from the surface during the time considered such as days. Thus, the constant value of soil temperature and water content at that depth can be used as lower boundary conditions T_n and θ_n of the equation (7.43) and (7.44). The upper boundary of soil temperature and soil water content for time interval $j+1$ is usually determined through considering a small change to the surface temperature and surface water content for time interval j.

7.4 Crank-Nicholson technique for the differential equations

The difficulty of solving surface energy balance model lies on the simultaneous solution of the two differential equations (7.43) and (7.44). Theoretical solution to the differential equations is extremely difficult due to many implicit relations. Usually the differential equations can be solved through explicit approximation method (Figure 7.2a) but the explicit method is only valid (*i.e.* convergent and stable) for $\delta t/\delta z^2 \leqslant 1/2$, in which δt is time interval and δz depth of soil layer (Smith, 1978). Considered the dimension of time in second and depth in meter, the explicit method will creates a giant

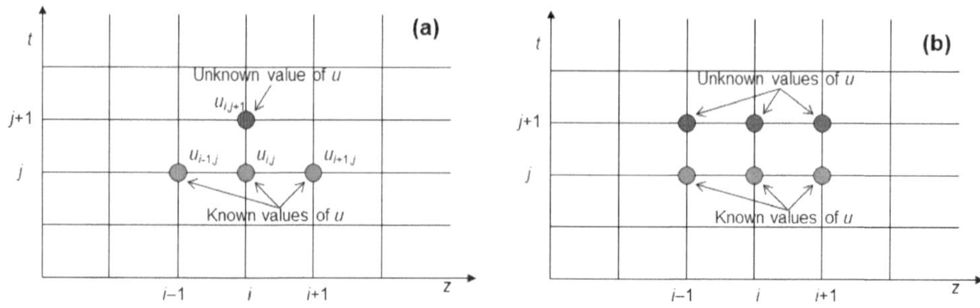

Figure 7.2 Numerical approximation to differential u by (a) explicit and (b) implicit methods.

computation volume for a short period of simulation such as one day. For example, if $\delta z = 0.1$ m, time interval has to be $\delta t \leqslant 0.005$ s in order to meet the convergent condition of the method and for simulating 1 s, it would take about 200 times of iterative computation.

Crank and Nicolson (1947) developed a method that reduces the total volume of calculation and is valid for all finite values of $\delta t/\delta z^2$ such as δt=60 s and δz=0.01 m. They approximated the difference by the mean of its finite-difference representations on the (j+1)th and the jth time rows (Figure 7.2b). Applying this implicit method, the differential terms of equation (7.43) can be approximated as follows.

$$C_h\partial h/\partial t=C_{hi+1/2,j}(h_{i,j+1}-h_{i,j})/\delta t \tag{7.47}$$

$$\partial(K_a\partial T/\partial z)/\partial z=\{K_{ai+1/2,j}[(T_{i+1,j+1}-T_{i,j+1})+(T_{i+1,j}-T_{i,j})]- K_{ai-1/2,j}[(T_{i,j+1}-T_{i-1,j+1})+(T_{i,j}-T_{i-1,j})]\}/2\delta z^2 \tag{7.48}$$

$$\partial(K_b\partial h/\partial z)/\partial z=\{K_{bi+1/2,j}[(h_{i+1,j+1}-h_{i,j+1})+(h_{i+1,j}-h_{i,j})]- K_{bi-1/2,j}[(h_{i,j+1}-h_{i-1,j+1})+(h_{i,j}-h_{i-1,j})]\}/2\delta z^2 \tag{7.49}$$

$$g\partial K_c/\partial z=g(K_{ci+1,j}-K_{ci,j})/\delta z \tag{7.50}$$

where $C_{hi+1/2,j}$, $K_{ai+1/2,j}$, $K_{ai-1/2,j}$, $K_{bi+1/2,j}$ and $K_{bi-1/2,j}$ are given as

$$C_{hi+1/2,j}=(C_{hi+1,j}+C_{hi,j})/2 \tag{7.51}$$

$$K_{ai+1/2,j}=(K_{ai+1,j}+K_{ai,j})/2 \qquad K_{bi+1/2,j}=(K_{bi+1,j}+K_{bi,j})/2 \tag{7.52}$$

$$K_{ai-1/2,j}=(K_{ai,j}+K_{ai-1,j})/2 \qquad K_{bi-1/2,j}=(K_{bi,j}+K_{bi-1,j})/2 \tag{7.53}$$

And according to equation (7.41), $K_{ai,j}$ and $K_{bi,j}$ are given by

$$K_{ai,j}=RK_{ci,j}\ln(h_{i,j})+h_{i,j}s_{i,j}D_{vi,j} \tag{7.54}$$

$$K_{bi,j}=RK_{ci,j}T_{i,j}/h_{i,j}+e_{vi,j}D_{vi,j} \tag{7.55}$$

Thus, equation (7.43) can be approximated as

$$C_{hi+1/2,j}(h_{i,j+1}-h_{i,j})/\delta t=\{K_{ai+1/2,j}[(T_{i+1,j+1}-T_{i,j+1})+(T_{i+1,j}-T_{i,j})]-K_{ai-1/2,j}[(T_{i,j+1}-T_{i-1,j+1})+ (T_{i,j}-T_{i-1,j})]+K_{bi+1/2,j}[(h_{i+1,j+1}-h_{i,j+1})+(h_{i+1,j}-h_{i,j})]- K_{bi-1/2,j}[(h_{i,j+1}-h_{i-1,j+1})+ (h_{i,j}-h_{i-1,j})]\}/2\delta z^2+ g(K_{ci+1,j}-K_{ci,j})/\delta z \tag{7.56}$$

Reorganization of the above equation leads to

$$-K_{ai-1/2,j}T_{i-1,j+1}+(K_{ai-1/2,j}+K_{ai+1/2,j})T_{i,j+1}-K_{ai+1/2,j}T_{i+1,j+1}-K_{bi-1/2,j}h_{i-1,j+1} +(K_{bi-1/2,j}+K_{bi+1/2,j}+2Z_tC_{hi+1/2,j})h_{i,j+1}-K_{bi+1/2,j}h_{i+1,j+1}= g_j \tag{7.57}$$

where $Z_t=\delta z^2/\delta t$ and g_j is defined as

$$g_j=2Z_tC_{hi+1/2,j}h_{i,j}+K_{ai+1/2,j}(T_{i+1,j}-T_{i,j})-K_{ai-1/2,j}(T_{i,j}-T_{i-1,j})+K_{bi+1/2,j}(h_{i+1,j}-h_{i,j})$$

$$-K_{bi\text{-}1/2,j}(h_{i,j}-h_{i\text{-}1,j})+2\delta zg(K_{ci+1,j}-K_{ci,j}) \tag{7.58}$$

Similarly, approximation of equation (7.44) leads to

$$C_{si+1/2,j}(T_{i,j+1}-T_{i,j})/\delta t=\{K_{di+1/2,j}[(T_{i+1,j+1}-T_{i,j+1})+(T_{i+1,j}-T_{i,j})]-K_{di\text{-}1/2,j}[(T_{i,j+1}-T_{i\text{-}1,j+1})+ (T_{i,j}-T_{i\text{-}1,j})]+K_{ei+1/2,j}[(h_{i+1,j+1}-h_{i,j+1})+(h_{i+1,j}-h_{i,j})]- K_{ei\text{-}1/2,j}[(h_{i,j+1}-h_{i\text{-}1,j+1})+(h_{i,j}-h_{i\text{-}1,j})]\}/2\delta z^2 \tag{7.59}$$

Reorganization of the above equation results in

$$-K_{di\text{-}1/2,j}T_{i\text{-}1,j+1}+(K_{di\text{-}1/2,j}+K_{di+1/2,j}+2Z_tC_{si+1/2,j})T_{i,j+1}-K_{di+1/2,j}T_{i+1,j+1} -K_{ei\text{-}1/2,j}h_{i\text{-}1,j+1}+(K_{ei\text{-}1/2,j}+K_{ei+1/2,j})h_{i,j+1}-K_{ei+1/2,j}h_{i+1,j+1}=G_j \tag{7.60}$$

where G_j, $K_{di+1/2,j}$, $K_{di\text{-}1/2,j}$, $K_{ei+1/2,j}$ and $K_{ei\text{-}1/2,j}$ are defined as

$$G_j=2Z_tC_{si+1/2,j}T_{i,j}+K_{di+1/2,j}(T_{i+1,j}-T_{i,j})-K_{di\text{-}1/2,j}(T_{i,j}-T_{i\text{-}1,j})+ K_{ei+1/2,j}(h_{i+1,j}-h_{i,j})-K_{ei\text{-}1/2,j}(h_{i,j}-h_{i\text{-}1,j}) \tag{7.61}$$

$$K_{di+1/2,j}=(K_{di+1,j}+K_{di,j})/2 \qquad K_{di\text{-}1/2,j}=(K_{di,j}+K_{di\text{-}1,j})/2 \tag{7.62}$$

$$K_{ei+1/2,j}=(K_{ei+1,j}+K_{ei,j})/2 \qquad K_{ei\text{-}1/2,j}=(K_{ei,j}+K_{ei\text{-}1,j})/2 \tag{7.63}$$

And $K_{di,j}$ and $K_{ei,j}$ are given as

$$K_{di,j}=K_{si,j}+h_{i,j}s_{i,j}LD_{vi,j} \tag{7.64}$$

$$K_{ei,j}=e_{vi,j}LD_{vi,j} \tag{7.65}$$

Using equations (7.47) and (7.60), one can simultaneously solved $T_{i+1,j+1}$, $T_{i,j+1}$, $T_{i\text{-}1,j+1}$, $h_{i+1,j+1}$, $h_{i,j+1}$ and $h_{i\text{-}1,j+1}$ when the initial values of these variables and boundary conditions are given. And this is the general case. The details of the solution to (7.57) and (7.60) are given in next section.

7.5 Gauss's method for solution of simultaneous equations

In equation (7.57) and (7.60), soil temperature T and relative humidity h for time interval j are known as initial conditions and the coefficients K_a, K_b, K_d and K_e can be computed for j. Thus, g_j and G_j are also known. For $j+1$, the upper boundary $T_{0,j+1}$ and $h_{0,j+1}$ and the bottom boundary $T_{n+1,j+1}$ and $h_{n+1,j+1}$ are also given. Therefore, the expansion of equation (7.57) about soil water movement in terms of relative humidity will result in the following simultaneous equations:

$$\begin{aligned} &+b_1T_1-c_1T_2+e_1h_1-f_1h_2=g_1 \\ &-a_2T_1+b_2T_2-c_2T_3-d_2h_1+e_2h_2-f_2h_3=g_2 \\ &\ldots\ldots \\ &-a_iT_{i\text{-}1}+b_iT_i-c_iT_{i+1}-d_ih_{i\text{-}1}+e_ih_i-f_ih_{i+1}=g_i \end{aligned} \tag{7.66}$$

$$\cdots\cdots$$

$$-a_nT_{n-1}+b_nT_n-d_nh_{n-1}+e_nh_n=g_n$$

where T represents soil temperature and h represent relative humidity in soil pore for time interval $j+1$. The subscript represents soil layer. The coefficients a, b, c, d, e, f, and g in all subscripts are known. Thus, the unknowns are T and h that need to be solved.

Similarly, equation (7.4.14) about soil temperature change can also be expanded into the following simultaneous equations:

$$\begin{aligned} &+B_1T_1-C_1T_2+E_1h_1-F_1h_2=G_1 \\ &-A_2T_1+B_2T_2-C_2T_3-D_2h_1+E_2h_2-F_2h_3=G_2 \\ &\cdots\cdots \\ &-A_iT_{i-1}+B_iT_i-C_iT_{i+1}-D_ih_{i-1}+E_ih_i-F_ih_{i+1}=G_i \\ &\cdots\cdots \\ &-A_nT_{n-1}+B_nT_n-D_nh_{n-1}+E_nh_n=G_n \end{aligned} \tag{7.67}$$

Totally, there are $2\times n$ unknown variables and $2\times n$ equations in (7.66) and (7.67). Thus, the unknown variables can be directly solved from the equation system by Gauss's elimination method.

To the simultaneous equations, the first equation in both (7.66) and (7.67) can be used to eliminate T_1 from the second one. Using the resulted equations from (7.66) and (7.67), h_1 can be eliminated to give a new second equation for it, which is in the same form as the first one. Similarly, using the first equation in (7.66) and the second equation in (7.67) and the second equation in (7.66) and the first equation in it, T_1 can be eliminated from both pair of equations. The resulted pair equation can be used to eliminate h_1 to give another new second equation in the same form as the first one for (7.67). And the new second equations can be used to eliminated T_2 and h_2 from the third equations and so on, until finally T_{n-1} and h_{n-1} are eliminated from the last equations in (7.66) and (7.67), giving two equations with only two unknowns, T_n and h_n. These two unknown can be easily solved from the two resulted equations. The unknowns T_{n-1}, T_{n-2}, …, T_2 and T_1 and h_{n-1}, h_{n-2}, …, h_2, and h_1 can then be found in turn by back-substitution.

Actually, this process can be simplified when the following substitution formula is applied for the solution of the simultaneous equations. Assume that the following stage of the elimination has been reached in (7.66) and (7.67),

$$a1_{i-1}T_{i-1}-b1_{i-1}T_i+d1_{i-1}h_{i-1}-e1_{i-1}h_i=g1_{i-1} \tag{7.68}$$

$$-a_iT_{i-1}+b_iT_i-c_iT_{i+1}-d_ih_{i-1}+e_ih_i-f_ih_{i+1}=g_i \tag{7.69}$$

$$A1_{i-1}T_{i-1}-B1_{i-1}T_i+D1_{i-1}h_{i-1}-E1_{i-1}h_i=G1_{i-1} \tag{7.70}$$

$$-A_iT_{i-1}+B_iT_i-C_iT_{i+1}-D_ih_{i-1}+E_ih_i-F_ih_{i+1}=G_i \tag{7.71}$$

where $a1_1=b_1$, $b1_1=c_1$, $d1_1=e_1$, $e1_1=f_1$ and $g1_1=g_1$ and $A1_1=B_1$, $B1_1=C_1$, $D1_1=E_1$, $E1_1=F_1$ and $G1_1=G_1$. Using (7.69)+(7.68)$a_i/a1_{i-1}$ to eliminate T_{i-1} leads to

$$(b_i-b1_{i-1}a_i/a1_{i-1})T_i-c_iT_{i+1}-(d_i-d1_{i-1}a_i/a1_{i-1})h_{i-1}$$
$$+(e_i-e1_{i-1}a_i/a1_{i-1})h_i-f_ih_{i+1}=g_i+g1_{i-1}a_i/a1_{i-1} \tag{7.72}$$

Similarly, eliminating T_{i-1} from (7.70) and (7.71) gives to

$$(B_i-B1_{i-1}A_i/A1_{i-1})T_i-C_iT_{i+1}-(D_i-D1_{i-1}A_i/A1_{i-1})h_{i-1}$$
$$+(E_i-E1_{i-1}A_i/A1_{i-1})h_i-F_ih_{i+1}=G_i+G1_{i-1}A_i/A1_{i-1} \tag{7.73}$$

For i=2, 3, …, we define

$$b1=b_i-b1_{i-1}a_i/a1_{i-1} \text{ and } B1=B_i-B1_{i-1}A_i/A1_{i-1}$$
$$d1=d_i-d1_{i-1}a_i/a1_{i-1} \text{ and } D1=D_i-D1_{i-1}A_i/A1_{i-1}$$
$$e1=e_i-e1_{i-1}a_i/a1_{i-1} \text{ and } E1=E_i-E1_{i-1}A_i/A1_{i-1}$$
$$g1=g_i+g1_{i-1}a_i/a1_{i-1} \text{ and } G1=G_i+G1_{i-1}A_i/A1_{i-1} \tag{7.74}$$

And substitute into equation (7.71) and (7.73), giving to,

$$b1T_i-c_iT_{i+1}-d1h_{i-1}+e1h_i-f_ih_{i+1}=g1 \tag{7.75}$$

$$B1T_i-C_iT_{i+1}-D1h_{i-1}+E1h_i-F_ih_{i+1}=G1 \tag{7.76}$$

Similarly, Using (7.78)+(7.68)$a1_i/A1_{i-1}$ to eliminate T_{i-1} leads to

$$(b_i-B1_{i-1}a1_i/A1_{i-1})T_i-C_iT_{i+1}-(d_i-D1_{i-1}a1_i/A1_{i-1})h_{i-1}$$
$$+(e_i-E1_{i-1}a1_i/A1_{i-1})h_i-F_ih_{i+1}=g_i+G1_{i-1}a1_i/A1_{i-1} \tag{7.77}$$

Eliminating T_{i-1} from (7.4) and (7.5) gives to

$$(B_i-b1_{i-1}A_i/a1_{i-1})T_i-c_iT_{i+1}-(D_i-d1_{i-1}A_i/a1_{i-1})h_{i-1}$$
$$+(E_i-e1_{i-1}A_i/a1_{i-1})h_i-F_ih_{i+1}=G_i+g1_{i-1}A_i/a1_{i-1} \tag{7.78}$$

For i=2, 3, …, let

$$b2=b_i-B1_{i-1}a1_i/A1_{i-1} \text{ and } B2=B_i-b1_{i-1}A_i/a1_{i-1}$$
$$d2=d_i-D1_{i-1}a1_i/A1_{i-1} \text{ and } D2=D_i-d1_{i-1}A_i/a1_{i-1}$$
$$e2=e_i-E1_{i-1}a1_i/A1_{i-1} \text{ and } E2=E_i-e1_{i-1}A_i/a1_{i-1}$$
$$g2=g_i+G1_{i-1}a1_i/A1_{i-1} \text{ and } G2=G_i+g1_{i-1}A_i/a1_{i-1} \tag{7.79}$$

And substitute into equation (7.77) and (7.78), giving to,

$$b2T_i - c_i T_{i+1} - d2h_{i-1} + e2h_i - f_i h_{i+1} = g2 \tag{7.80}$$

$$B2T_i - C_i T_{i+1} - D2h_{i-1} + E2h_i - F_i h_{i+1} = G2 \tag{7.81}$$

Using (7.75)−(7.76)$d1/D1$ to eliminate h_{i-1} leads to

$$(b1 - B1d1/D1)T_i - (c_i - C_i d1/D1)T_{i+1} + (e1 - E1d1/D1)h_i$$
$$-(f_i - F_i d1/D1)h_{i+1} = g1 - G1d1/D1 \tag{7.82}$$

And eliminating h_{i-1} from (7.80) and (7.81), we also get

$$(B2 - b2D2/d2)T_i - (C_i - c_i D2/d2)T_{i+1} + (E2 - e2D2/d2)h_i$$
$$-(F_i - f_i D2/d2)h_{i+1} = G2 - g2D2/d2 \tag{7.83}$$

Defining

$$\begin{aligned} &a1_i = b1 - B1d1/D1 \text{ and } A1_i = B2 - b2D2/d2 \\ &b1_i = c_i - C_i d1/D1 \text{ and } B1_i = C_i - c_i D2/d2 \\ &d1_i = e1 - E1d1/D1 \text{ and } D1_i = E2 - e2D2/d2 \\ &e1_i = f_i - F_i d1/D1 \text{ and } E1_i = F_i - f_i D2/d2 \\ &g1_i = g1 + G1d1/D1 \text{ and } G1_i = G2 - g2D2/d2 \end{aligned} \tag{7.84}$$

And substituting into (7.82) and (7.83), we get

$$a1_i T_i - b1_i T_{i+1} + d1_i h_i - e1_i h_{i+1} = g1_i \tag{7.85}$$

$$A1_i T_i - B1_i T_{i+1} + D1_i h_i - E1_i h_{i+1} = G1_i \tag{7.86}$$

These two equations are exactly in the same form as those in (7.68) and (7.70). Thus, they can be used for the next elimination.

After eliminating n times, we reach the final two pairs of equation as follow

$$a1_n T_n + d1_n h_n - g1_n \tag{7.87}$$

$$A1_n T_n + D1_n h_n = G1_n \tag{7.88}$$

Using (7.87)$A1_n$−(7.88)$a1_n$ to eliminate T_n, we get the solution of h_n as

$$h_n = (g1_n A1_n - G1_n a1_n)/(d1_n A1_n - D1_n a1_n) \tag{7.89}$$

Similarly, the solution of T_n is

$$T_n = (g1_n D1_n - G1_n d1_n)/(a1_n D1_n - A1_n d1_n) \tag{7.90}$$

After getting the solution of T_n and h_n, we can get all other unknowns from equation (7.85) and (7.86) for $i=n-1, n-2, \cdots, 1$. For convenience of derivation, we define

$$g3 = g1_i + b1_i T_{i+1} + e1_i h_{i+1} \tag{7.91}$$

$$G3 = G1_i + B1_i T_{i+1} + E1_i h_{i+1} \tag{7.92}$$

in which h_{i+1} and T_{i+1} are known from (7.89) and (7.90) when $i+1=n$, and rewrite the equations (7.85) and (7.86) as

$$a1_iT_i+d1_ih_i=g3 \tag{7.93}$$

$$A1_iT_i+D1_ih_i=G3 \tag{7.94}$$

Using (7.93)$A1_i$−(7.94)$a1_i$ to eliminate T_i, we get the solution of h_i as

$$h_i=(g3A1_i-G3a1_i)/(d1_iA1_i-D1_ia1_i) \tag{7.95}$$

Similarly, the solution of T_i is

$$T_i=(g3a1_i-G3A1_i)/(a1_iD1_i-A1_id1_i) \tag{7.96}$$

Continuing the above procedure, we can get the solution of all unknowns.

7.6 Computation procedure of the model

As described above, the numerical approximation for the solution of surface energy balance model involves the re-adjustment of some given parameters in the iterative calculation. The computation actually is rather complicated. The procedure for the numerical solution of the model can be summarized as follows:

(1) Read required constants and coefficients.

(2) Assume initial soil temperature and water content for each layer.

(3) Compute initial soil relative humidity for each layer.

(4) Compute the required soil parameters: $D_v, e_v, s, C_h, K_c, C_s, K_s$.

(5) Input data of global radiation, air temperature, air relative humidity and wind speed: R_s, T_a, h_a, u_z.

(6) Give a start point of surface temperature T_s.

(7) Compute net radiation R_n and sensible heat H.

(8) Give a start point of surface relative humidity h_s.

(9) Calculate surface vapor pressure and compute LE by equation (7.9).

(10) Solve soil temperature T_i and relative humidity h_i from equation (7.43) and (7.44) by Crank-Nicholson technique.

(11) Calculate LE_c by equation (7.12) and compute $f(h_s)$ by equation (7.33).

(12) If $f(h_s)$ is small enough to reach the required accuracy, then go to step (15).

(13) Get a small δh and follow step (9) to (11) to calculate $f(h_s+\delta h)$.

(14) Update h_s in the form of equation (7.32) and repeat step (9) to (12).

(15) Compute soil heat G by equation (7.16).

(16) Compute $f(T_s)$ by equation (7.29).

(17) If $f(T_s)$ is small enough to reach the required accuracy, then go to (20).

(18) Get a small δT and follow step (7) to (16) to calculate $f(T_s+\delta T)$.

(19) Update T_s by equation (7.32) and repeat step (7) to (17).

(20) Output the results for the time interval: R_n, H, LE, G, T_s, h_s and so on.

(21) Repeat step (4) to (20) until terminal of the simulation.

The procedure involves several iterations and the actual computation, as we can imagine, is very complicated. Using Quick BASIC 4.5, we have programmed the model for practical application (see Appendix B). The main flow chart of the computation of the model is illustrated in Figure 7.3 and the flow chart of computing $f(T_s)$ and $f(T_s+\delta T)$ in Figure 7.4.

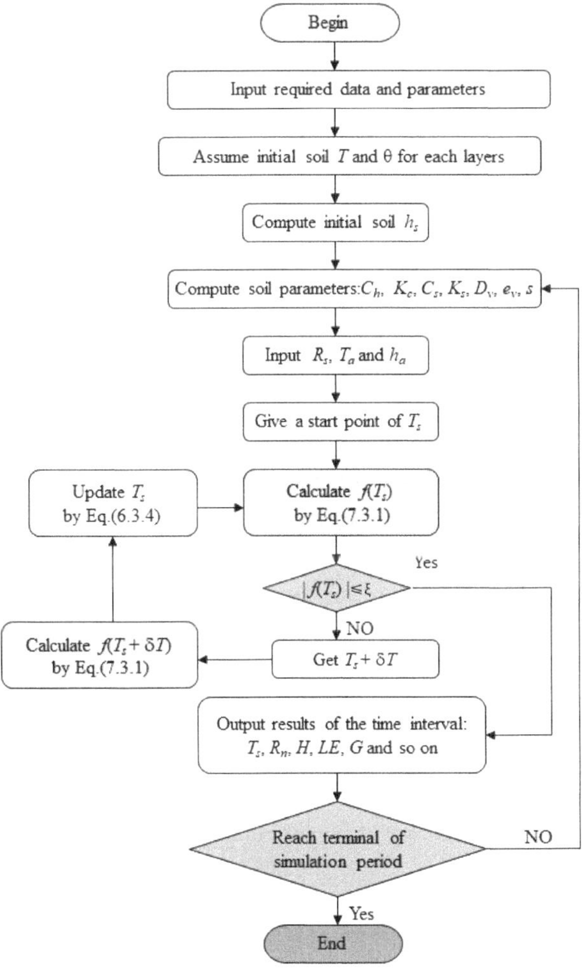

Figure 7.3 The main flow chart of the computation of the model. ξ is the required accuracy for the computation

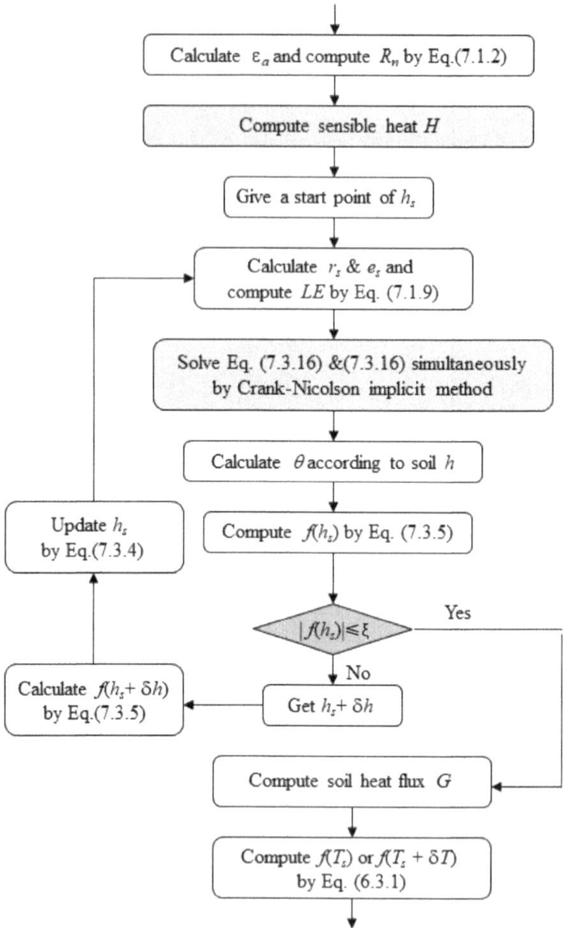

Figure 7.4 The flow chart of computing $f(T_s)$ and $f(T_s+\delta T)$. ξ is the required accuracy for the computation

7.7 Validation of the model through application to south Israeli desert

In order to validate the model, I applied it to the south Israeli desert for comparison of soil temperature at various depths, using the meteorological data from Sede Boker Meteorological Observation Station. The Station locates in the center of an alluvial plain with about 2km wide stretching from east to west. In the north of the plain are the low hills with about 20m height above the plain and in the south is a long crater with about 50m depth and km width. The data of July 16, 1998 was selected for the validation of the model. It is the hot dry season of the region and the dynamics of micrometeorological events is subjected to local conditions. The sky on July 16 was very clear and it represented the general case of the season. The soil is gray alluvium mainly composed of silt and clay. Albedo of the soil is about 30% and volumetric soil water content ranges from about 85kg/m^3 at the surface to 120kg/m^3 at 50cm depth.

Comparison of lw simulated to measured the soil temperature at 10cm and 30cm depths is shown in Figure 7.5a. The measured soil temperature change shown in Figure 7.5 is a typical one in south Israeli desert. A similar change was found in Fania and Zipora (1997) and Zemel and Lomas (1977). The very close change of simulated soil temperature to the measured one in Figure 7.5a proofs the high validity of the model. The average difference of simulated and measured soil temperature is within

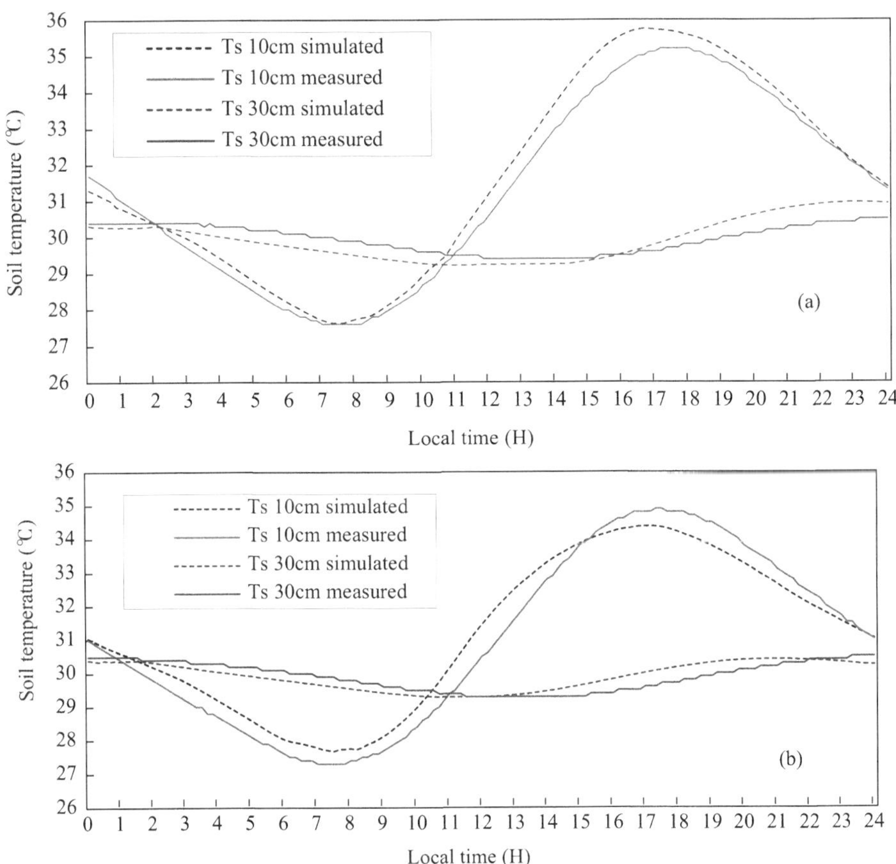

Figure 7.5 Validation of the simulation model through application to the south Israeli desert, illustrating the comparison of simulated and measured soil temperatures at 10cm and 30cm depth for data sets of July 16, 1998 (a) and July 5, 1997 (b).

0.4°C with maximum of 0.95°C at 10cm depth and within 0.3°C with maximum of 0.53°C at 30cm. The ratio of the difference in the daily vibration is usually used to evaluate the accuracy of simulation. In our case, the daily vibration of soil temperature was about 7°C at 10cm depth. Thus, the ratio is about 5.71%, which is quite small, hence indicates a quite accurate simulation.

The model was also applied to the data set of 1997 in the same season for mutual validation. The simulated soil temperature at 10cm depth also has a good matching with the measured one (Figure 7.5b). Due to the difference in air temperature and air humidity, the accuracy of soil temperature matching for the data set is with an average difference of about 0.6°C with maximum of 1.2°C at 10cm depth. The daily vibration of soil temperature was about 8°C. Thus, the ratio of the difference in the daily vibration is about 7.50%, which is also quite small. Therefore, it can be concluded that the simulation results are very close to what actually happen in the arid alluvial plain.

The successfulness of applying the model to estimate heat fluxes and surface temperature for micrometeorological analysis relies on the successful determination of required soil parameters. It is very important to reasonably determine the parameters in equation (7.19) or (7.20), (7.21) and (7.24) for computing soil water potential, hydraulic conductivity and thermal conductivity and the roughness for computing air resistance to heat transfer. Besides, a proper way of operating the simulation procedure is also very important for numerical solution of the model. The above simulation employs 1 minute for time interval. For the soil profile, a 0.5m depth was considered because daily soil heat penetration in the region is limited within this depth (Fania 1997). The profile was divided into 20 layers. Time interval greater than 5 minutes and soil layer thicker than 0.1m may not be very good for an accurate simulation. And usually it is better to run the model for 2-3 days before the simulation results for the destination day is outputted.

7.8 Summary of the chapter

A micrometeorological model has been establelished from the viewpoint of surface energy balance in the land-atmosphere interface system for simulating the surface temperature change and heat flux variation of the arid region. The model couples soil temperature change with soil moisture movement for estimation of soil heat flux and latent heat flux. A complete description of the model is presented in the chapter. Through the two essential factors, *i.e.* surface temperature and surface moisture change, the model can be numerically solved for various study purposes such as irrigation program and micrometeorological analysis. The numerical solution of the model involves the estimation of the dynamics of heat fluxes and many useful soil-water and meteorological parameters required for earth resource management.

A methodology of numerical solution of the model is presented with details so that it can be programmed for its application to the real world. Soil and latent heat fluxes are determined by soil temperature change and soil moisture movement, which can be described as differential equations. Crank-Nicolson implicit method is used to expand the differential equations into two sets of simultaneous linear equations, which are then solved by applying Gauss's elimination method. The successful application of the two critical techniques is the basis for the numerical solution.

The change of soil moisture in the time interval is equal to the evaporation from the ground surface within the interval. Newton-Raphson approximation method is used for the iterative computation of solving latent heat flux to meet this requirement. This approximation method is also applied to the solution of surface temperature of the model. Because the method is rather complicated, a detailed computational procedure of its numerical solution is also presented. Using this procedure, heat fluxes and temperature change can be easily estimated from the model when the required soil parameters and meteorological data are available. And this is the general case. Therefore, the methodology presented herein provides an easy way of applying surface energy model to simulate the dynamics of many micro-meteorological phenomena in the soil-air interface.

Using the meteorological data from Sede Boker in south Israeli desert, the model and its numerical solution method have been validated through comparison of soil temperature at 10cm and 30cm depths. The difference between the measured and the simulated soil temperatures is within 0.6℃ at 10cm depth. The ratio of the difference to the daily vibration of soil temperature is less than 10%, which is quite small, hence indicates a quite accurate simulation. Good matching of the simulated soil temperature to the measured one proofs the validity of the model and its numerical solution method. Except the importance of determining the required soil parameters, successful application of the model and its numerical solution also depends on the proper way of operating the computation procedure.

8 Preparation of data for simulation

In order to apply the model to simulate LST change on both sides of the region, data of the functioning factors are required to prepare. These include global radiation, surface albedo, soil water content, soil porosity and density, soil constituents, soil temperature, air temperature, relative humidity of the air and wind speed near the surface.

8.1 Global radiation of the region

Global radiation is the basis of surface energy balance in the region because there are no other sources from the interior earth. Assumed solar radiation being constant at the top of the atmosphere, several proposed methods are available to determine the global radiation on the ground (Brutsaert, 1982; Davies and Idso, 1979; Monteith, 1973). However, when suitable instrument is available, direct measurement is the common way of determining global radiation for specific study.

The observed LST anomaly on both sides appears the sharpest in summer, which is dry and hot. This implies that the study should focus on the season. The measurement of global radiation is selected to be in July 16-20, 1998, when the sky was very clear and the region was very dry and the vegetation was dormant. A pyranometer operating in the spectrum ranging from ultraviolet through visible to near infrared wavelength was used for the measurement. The pyranometer was mounted upward on top of a frame above 1 meter from ground, which was carefully installed so that it was in the water level status to record the radiation from hemispheric direction. A datalog and a storage module were used to record the output from the pyranometer in a frequency of every 5 seconds for one record. The raw data was averaged to one record per minute after download into computer for further calculation.

The measuring results are shown in Figure 8.1, which indicates that the global radiation during the measuring period is very similar with each other. The shape of the radiation curves is highly similar with a sine one. The curves start from a little bit earlier than local time 6:00 in dawn and end at 19:45 in dusk, representing the time span for the region to receive the incoming radiation for its surface energy balance process. From 6:00 in morning, the radiation rapidly increases with time, due to the increase of the sun's shining angle. At about 13:00 when the sun's ray is about perpendicular to the ground, the radiation curves reach their apex and from this point to about 17:45 in dusk, the radiation declines gradually. At the apex hour 13:00, global radiation of the region is about 1130 W/m^2 in average, with the lowest of 1100 W/m^2 and the highest 1150 W/m^2 during the measuring period (Figure 8.1). Another feature is that the fluctuation in morning is much less than in afternoon. During the measuring period, daily change of the radiation is less than 15 W/m^2 from 6:00am to 11:00am. However, the change in afternoon is above 35 W/m^2. For a specific day, the fluctuation is much more obvious in the noon hours than others (Figure 8.1). After average, the curve is very smooth in morning and afternoon except a little bit fluctuation in 12:30-13:30. The average global radiation will be used in the modeling.

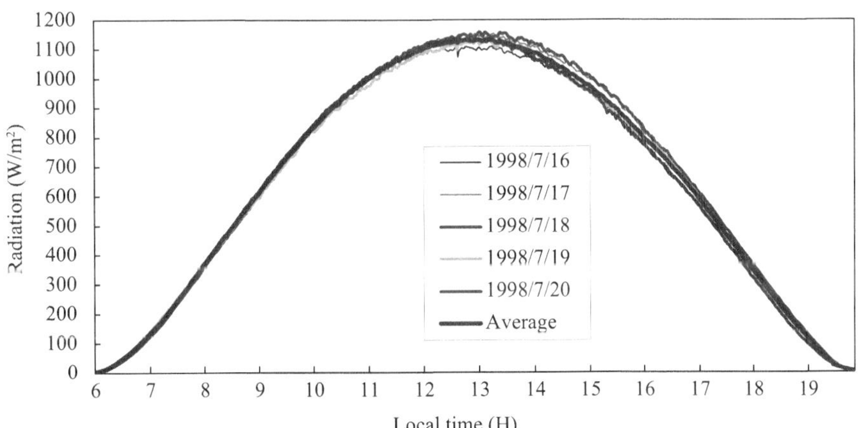

Figure 8.1 Global radiation of the region with vibration during the measuring period and their average.

8.2 Albedo of the surface patterns

At the interface of the soil-atmosphere system, part of the incident global radiation is reflected by the ground surface into the atmosphere. Only part of the incident radiation participates in the process of thermal exchange at the interface. The albedo of the surface determines how much incident global radiation is reflected back into the atmosphere. Generally speaking, the brighter the surface, the higher its albedo is. And the higher the albedo, the less the net radiation for a given global radiation. This means that the darker surface will absorb more incoming energy than the brighter one. As mentioned in chapter one, the Israeli side has more vegetation and biogenic crust hence behaves as darker than the Egyptian side. Therefore, it can be logically referred that the Israeli side absorbs more incident energy. Does this means that the observed higher LST on the Israeli side is directly resulted from its more absorption of the incident radiation? This can only be answered through the simulation of energy balance at the interface of soil-atmosphere system.

8.2.1 Measurement of incident and reflected radiation

Albedo is defined as the ratio of reflected radiation to the incident radiation. In order to calculate the albedo of ground surface, we need to measure the reflected radiation and the incident one simultaneously. The two pyranometers were mounted at a frame about 1 meters from the ground. One pyranometer was put upward to measure the incident sky radiation and the other downward to acquire the reflected radiation from ground. Both the pyranometers were kept at the water level. In order to minimize the impact of sky radiation from unexpected directions, a panel was also mounted at the bottom of the pyranometers to force it only recording the radiation from a hemispheric direction. The measuring period was the same as radiation measurements. Before calculating the albedo, the data from the two pyranometers were calibrated into a common level. In practice, the pyranometer recording the incident radiation was used as the base to calibrate the pyranometer recording the reflected radiation.

8.2.2 Albedo of the typical surface patterns

The average albedo of the typical surfaces is shown in Figure 8.2a. One obvious common feature of the albedo change is that its amount declines with the angle of the sun's ray increase. Usually the sun's ray is perpendicular to the ground at about noon (13:00 local time). Therefore, the albedo of the surfaces is minimal around this time (13:00-14:00). Actually, The NOAA-AVHRR also passes the region at about 14:00. That is, when the NOAA-AVHRR remotely scans the region, surface albedo is at the lowest level. The albedo of all surface patterns is extremely high at dawn and dusk. At this time, the azimuth angle of the sun's ray is very small and the reflected fraction of the incident radiation is very high, albeit both incident and reflected radiations are very small (<50 W/m^2).

The albedo of the four surface patterns also has great difference. Sand has the highest albedo among the four. At about 8:00 in morning, its albedo is about 60%. The albedo gradually decreases to 46% at about 13:00-14:00. In afternoon, it experiences a gradual increase to about 55% at about 19:00. Between 19:00 to 19:30, the albedo rapidly increases from about 55% to 100%. The change of the albedo in morning and afternoon is not symmetrical. During the five hours in the morning from 8:00 to 13:00, the albedo changes from about 60% to 46%, with 13% decrease. In the afternoon from 14:00 to 19:00, it changes from 46% to 55%, with only 9% increase, which is 4% less than the change in the morning.

Biogenic crust has much lower surface albedo than sand during the day from 8:00 to 19:00. In the morning at about 8:00, the albedo of biogenic crust is about 40%, which is 20% less than that of sand. The albedo gradually decreases to 33% at the hour from 13:00 to 14:00 when the sun is at zenith. From this minimum, it experiences a gradual increase to about 40% at 19:00. This change process is much more symmetrical than that of sand. Albedo difference between biogenic crust and sand is very obvious during the hour 13:00-14:00. Sand's albedo is about 13% higher than that of biogenic crust. This difference is very important in leading to the LST difference on both sides, because the Israeli side is mainly occupied by biogenic crust while the Egyptian side by sand. The albedo difference between the two surfaces still remains at about 13% in afternoon till 19:00.

Shrub is the main vegetation in summer when annuals die. Thus the measurement is done on a shrub with a big canopy for calculating the albedo of vegetation. Figure 8.2a indicates that, during daytime from 7:00 to 18:00, the albedo of vegetation is the lowest among the four. The albedo curve of vegetation reaches the lowest (about 20%) at 12:00-13:00. However, it still remains at the minimal level at 13:00-14:00. The albedo has very small changes (<2%) in afternoon from 14:00 to 16:00. This is obviously different from other surface patterns, which experience much faster increase of albedo during the hours. This is mainly because the shrub's chlorophyll has strong absorption to the incoming radiation even the plant is in dormancy.

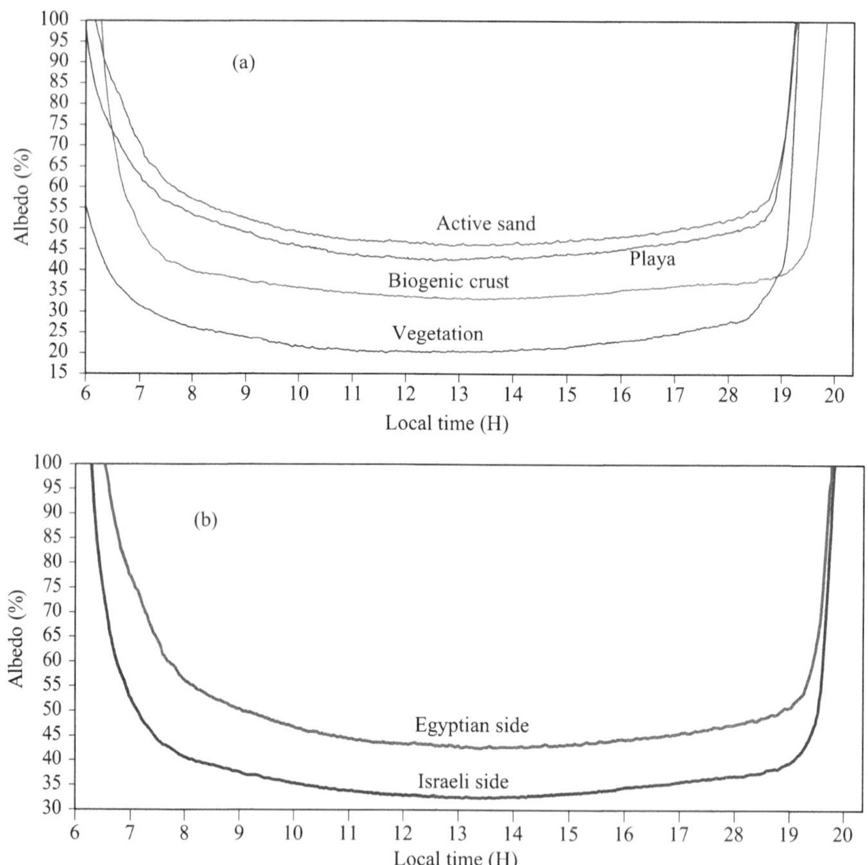

Figure 8.2 Albedo of main surface patterns (a) and average albedo of Israeli and Egyptian side (b).

As shown in Figure 8.2a, playa has very similar albedo with sand. The albedo of playa is also very high. Actually, its albedo is only a little bit (about 2%-5%) lower than that of sand. The albedo of playa changes from about 75% in the morning at 7:00 to 55% at 8:00. The rapid decrease continues in the following hours. At 9:00, the albedo decreases to 50% and at 10:00 it is 47%. The albedo reaches the valley of about 43% at about 13:00, then it gradually increases in the following hours. At 19:00, the albedo resumes to 50%. The increase rate of the albedo in the afternoon is less than its decrease rate in the morning. Another feature is that its minimum hour is not exactly, like other surface patterns, between 13:00 and 14:00 but between 12:30 and 13:30.

8.2.3 Albedo of ground surface on both sides

The surface composition has been used to estimate the average surface albedo on both sides. The result is shown in Figure 8.2b. Due to the higher percentage of biogenic crust and vegetation, which have lower albedo, the Israeli side has much lower surface albedo. At about noon, the average albedo is about 32% on the Israeli side and about 44% on the Egyptian side.

Except the numerical difference, similar feature of albedo change on both sides can be seen. Surface albedo is extremely high in both dawn and dusk. At about 6:30 in the morning, the albedo of the Egyptian side is high up to about 100% while that of Israeli side is only about 75%. This means that the Israeli side absorbs more incident radiation in dawn even the absolute amount is very low ($<50\,W/m^2$) and that the time for the Israeli side to start reflecting the incoming radiation is a little bit later (20minutes) than the Egyptian side. However, the situation in dusk is similar with each other. The albedo on both sides experiences a rapid decent in early morning and a rapid ascent in late afternoon. At about 7:30 local time in the morning, the Israeli side has albedo of about 40%-45%, about 20% less than the Egyptian side. At about 10:00, the albedo of the Israeli side drops to about 35% and the Egyptian side 48%. The difference between the two sides continues to narrow in the following hours till it reaches the lowest point at noon when the difference is about 12%. It can be rationally expected that the albedo difference have great effect on the surface temperature change and its difference on both sides.

8.3 Soil water content

Water content of soil profile is needed for the modeling in two ways: (1) determination of initial and boundary conditions for the solution of differential equations about soil water movement and temperature change, (2) validation of simulation result about soil water movement in the region. Generally, soil water content can be determined in several methods (Kutilek and Nielson, 1994; Gardner et al., 1991; Ghildyal and Tripathi, 1987). When suitable instruments such as neutron moisture meter are available, soil water content can be easily and accurately measured in the field by installing the probes at various depths of soil profile (Marshall and Holmes, 1979). For the study, I lack of the proper instrument for direct measurement. Thus, I have to use the classic gravimetric method of drying soil samples for the measurement. Actually soil water percolation and movement of the sand dunes had been studied in the research of Yair et al. (1997), who used neutron probes to monitor soil water content change for the analysis in the rainfall years 1991-1992 and 1992-1993. Soil water content data of the region in recent years are not available and the comparison of soil water content under sand and biogenic crust surfaces also still remains as undone.

The samples for the measurement of soil water content in my current study were taken from the field at Nizzana Research Site (see Figure 5.4). A series of sampling have been carried out for the measurement in various seasons. The results are presented as follows.

8.3.1 Soil water content of surface layer in dry season

The first sampling date was on August 1, 1997. The surface of the region was extremely dry because it had no rain for about four months. Four metal cans were used to hit into soil from surface for the sampling. Two cans were for sand surface and two for biogenic crust surface. The depth of the cans is 11cm with diameter 7.3cm. Thus, the volume is about 460cm^3. The can weights about 65g. The soil samples were immediately wrapped in the field using plastic containers to prevent moisture escape before weighting and put into a large box to avoid direct sunshine. An accurate electronic scale was used to weight the soil samples before putting them into an oven of 105°C for drying soil moisture out of the samples. After baking for 24 hours, the samples were weighted again to get the net gross weight. Then, they are subtracted the can weight to get the net soil weight. The water content was calculated from the weight different of the soil samples before and after drying. The result is shown in Table 8.1.

Table 8.1 Soil water content of surface layer (11cm depth), Aug. 1, 1997

Soil samples	Total (g)	Soil (g)	Water (g)	Soil water content (%)	
				Mass	Volume
Sand 1	695.70	694.00	1.70	0.24436	0.4016
Sand 2	689.20	687.50	1.70	0.24667	0.4016
Average	692.45	690.75	1.70	0.24551	0.4016
Biogenic crust 1	745.20	742.90	2.30	0.30864	0.5433
Biogenic crust 2	736.30	733.50	2.80	0.38028	0.6614
Average	740.75	738.20	2.55	0.34425	0.6024

This result represents the surface water content of the region in dry season even though the sampling depth is about 11 cm. Table 8.1 indicates that the surface of the region is extremely dry. With the depth of 11 cm from the surface, the top layer of soil profile only has volumetric soil water content of about 0.40% for sand surface and 0.60% for biogenic crust. In mass dimension, the percentage is even less. Table 8.1 also indicates an interesting phenomenon. Biogenic crust has relatively higher moisture content than sand surface. In mass dimension, the moisture content in biogenic crust is about 0.1% higher than in sand surface. In volumetric dimension the difference is about 0.2%. This maybe implies that soil moisture content in biogenic crust is really higher than in sand surface.

8.3.2 Change of soil water content after a heavy rain process

The second sampling date was on January 16, 1998, which represents the typical situation of wet season. There was a heavy raining process in January 11-12, 1998, with a total precipitation of 23.9mm, one of the two maximal raining processes of the wet season. Therefore, the sampling date was just about 5 days after the heavy rain. Both sand and biogenic crust surfaces in the field were obviously wet. The fungi and other microphytes developed rampantly on biogenic crust. Digging a profile on top dune found that the top wet layer was only about 40cm depth and the soil under this depth still remained in relatively dryer. This indicated that the soil water under sand surface had not yet penetrated into the lower layer. The sand surface was relatively dryer than the beneath layer. In order to compare the impact of sampling depth on soil water content, both cans and plates were used to take soil samples from sand and biogenic crust surfaces. The plate has a depth of 1.4 cm with a diameter of 8cm and a weight of 18g. Consequently, the volume is about 70.4cm^3. The measurement results are listed in Table 8.2.

Three features can be seen from Table 8.2. First of all, soil water content after heavy rain is very high. Because heavy rain is rare in the desert region, this measurement may represent the maximal surface water content of the region in wet season even it is far from saturation. Second, sampling depth has strong effect on soil water content. This is naturally explicit. Due to the exposure to the atmosphere, the top layer is always drier than the adjacent beneath layer. This is why the plate samples have less water content than the can ones. Because the biogenic crust is only a few millimeters (1-5mm) in thickness, the data from plate samples would be closer to the true difference of water content between biogenic crust and sand surface. Third, the biogenic crust again has higher water content than sand in both can and plate samples. This result simply validates what was observed in the first sampling in dry season.

Soil samples were taken not only from the surface layer but also from various depths of soil profile on sand surface for measuring the moisture content of the profile. Except the surface layer represented by plate samples, soil samples from five depths were also measured for the moisture content. Each layer is represented by three individual samples. The average water content of the sand profile is shown in Figure 8.3a. Because the penetration of soil water from top to bottom is a slow process, the moisture content at the depth of about 50cm from surface is much lower than the layers above it. As indicated above, the heavy rain process provided sufficient water supply for the top layers but the water still remains in the layers near the surface and has not yet penetrated into the lower layer. Thus, when I dug the soil profile, I found that the soil at about 40cm was obviously dry by eye observation. And only the soil at above 40cm depth was seen to be wet due to the rain.

Table 8.2 Soil water content of surface layer, January 16, 1998

Soil samples	Total (g)	Soil (g)	Water (g)	Soil water content (%)	
				Mass	Volume
For the can samples (11cm depth)					
Sand 1	751.10	700.90	50.20	6.6835	11.8584
Sand 2	733.60	683.50	50.10	6.8293	11.8347
Sand 3	745.00	692.70	47.30	5.3489	11.1733
Average	741.57	692.37	49.20	6.6346	11.6221
B. crust 1	789.80	670.50	119.30	15.1051	28.1813
B. crust 2	782.70	658.90	123.80	15.8170	29.2443
B. crust 3	786.80	665.50	121.30	15.4169	28.6538
Average	786.10	664.97	121.47	15.4518	28.6931
For the plate samples (1.4cm depth)					
Sand 1	164.18	156.89	7.29	4.4402	10.4143
Sand 1	164.18	156.89	7.29	4.4402	10.4143
Sand 2	156.74	149.14	7.60	4.8488	10.8571
Sand 3	161.39	154.84	6.55	4.0585	9.3571

(续表)

Soil samples	Total (g)	Soil (g)	Water (g)	Soil water content (%)	
				Mass	Volume
Average	160.77	153.62	7.15	4.4453	10.2095
B. crust 1	133.84	121.58	12.26	9.1602	17.5143
B. crust 2	129.43	117.58	11.85	9.1555	16.9286
B. crust 3	135.02	120.39	14.63	10.8354	20.9000
B. crust 4	120.38	106.92	13.46	11.1933	19.2286
B. crust 5	129.25	117.67	11.58	8.9594	16.5429
B. crust 6	126.12	112.13	13.99	11.0926	19.9857
Average	129.01	116.05	12.96	10.0473	18.5167

Another important feature shown in Figure 8.3a is that the soil moisture at layers below 50cm is very low with little variation. Because the soil water supplied by the heavy rain is still hold in the upper layers, the lower layers still remain at its dry condition of low moisture content. At about 1m from the surface, the volumetric moisture content is about 5.5%. This value may represent the constant soil water content at dry season.

The third sampling date was on January 25, 1998. Since the last sampling on January 16, there was no rain. Field observation revealed that surface became dry again, especially on sand surface. Digging soil profile on sand surface found that the abrupt distinguish between upper wet and lower dry layers disappeared and the soil wetness seemed to be gradually changing from top to bottom of the profile. The plates with the same sizes were used to take the soil samples from the field for the measurement. The result is also plotted in Figure 8.3a for comparison.

Soil water content differs greatly between sand and biogenic crust, with the later obviously higher. In average, the surface layer of sand has volumetric soil water content of 7.01%, which is about 5% less than that of biogenic crust (Figure 8.3a). This is probably due to the structural difference of these two surfaces in the capacity of soil water keeping. It has been well known the water penetration in sand is much faster than in other kind of natural soil (Ghildyal and Tripathi, 1990; Kutilek and Nielson, 1994). Other reasons may include the difference of evaporation and spatial location. In a short time after rain, biogenic crust may have a function of slowing down evaporation and preventing soil water lose from the surface through blocking the vapor flowing into the air. Usually sand surface locates in relatively high places such as the top dunes and the biogenic crust in low places such as interdunes. Naturally water is driven by gravity to flow from high to low even though runoff in the arid region is few. The difference of spatial location may lead to the difference of soil water supply on the two kinds of surfaces and finally this difference probably makes the water content measurements based on the sampling be naturally different.

One eminent feature that can be drawn from the result of sampling on January 25 is that volumetric water content in layers at various depths is much higher under biogenic crust than sand surface. At about half meter from the surface, soil water content is about 18.34% for the profile under biogenic crust and about 9.37% under sand surface. The former is almost double the later. This represents the maximal soil water content of the region after a heavy raining process. This feature can also be seen at the depth of about 80cm. However, the difference of soil water content at about 100cm is much less than the layers above it. Under biogenic crust, the content is about 7% and under sand surface about 6%.

Compared with January 16, one can see that soil water has penetrated from upper into lower layers. The surface and the layer at about 10cm depth have very high soil water content on January 16. Due to surface evaporation and penetration, the two layers have lower water content on January 25. However, the content in the layers below 30cm increases due to the water supply from upper layers through penetration. Obvious increase of soil water content can be seen at about 50cm.

8.3.3 Behavior of soil water content change in wet season

In arid environment, soil experiences a rapidly drying process after rain. Usually soil surface resumes to very dry status if there is no rain in about 3-4 weeks. This can be seen from the measurement on February 3, 1998 (Figure 8.3b). Field observation on February 3 found that the ground was much dryer than on January 25. However, digging soil profile found that soil layer at about 10cm still remains to be obvious wet by eye observation. Soil samples from profiles were taken in plates for measuring their water content. In order to have a better observation about soil water content change at various depths, the profiles were dug up to 1.2m depth.

Compared with the last sampling on January 25, the surface on February 3 was much drier than the layers beneath it on both biogenic crust and sand. In average, sand surface only has volumetric water content of 0.35%, which is about 20 times less than last sampling day. The soil water content of surface layer on biogenic crust also drops from 11.84% to only 0.85%. This implies that the region experienced a dramatically fast drying process in a short period.

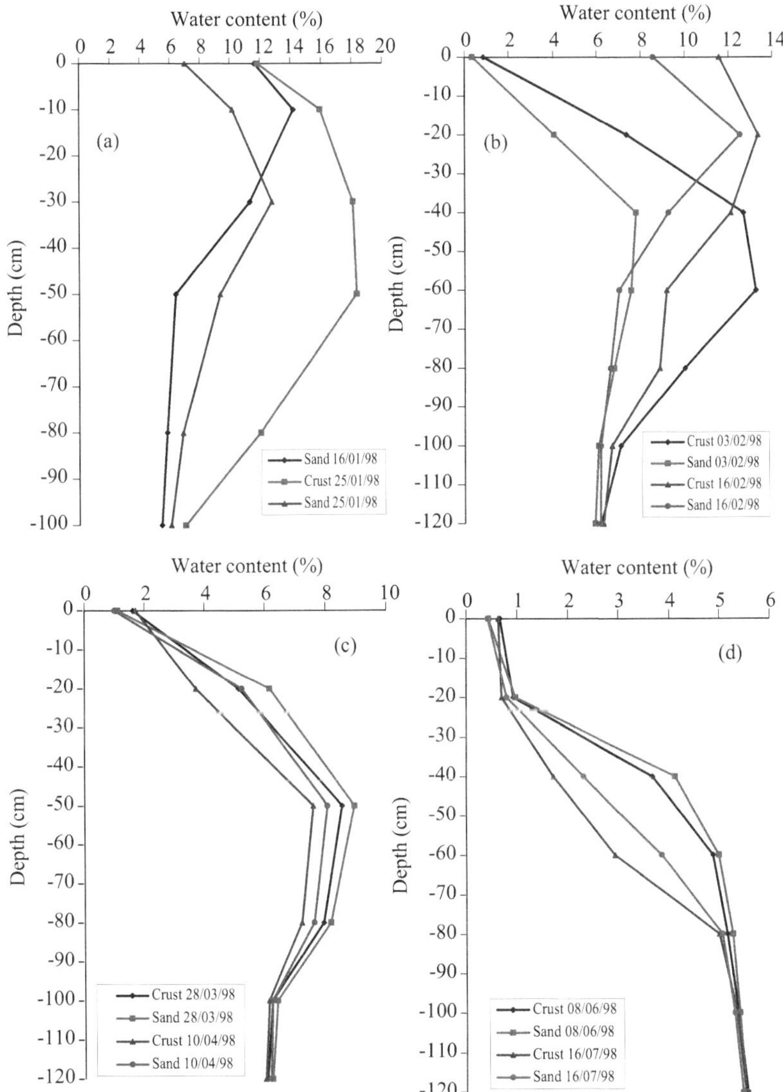

Figure 8.3 Volumetric soil water content at various depths of profiles under sand and biogenic crust surfaces, illustrating the results based on the sampling date of (a) January 16 and 25, 1998; (b) February 3 and 16, 1998; (c) March 28 and April 10, 1998 and (d) June 8 and July 16, 1998.

At the same time, soil profiles under these two surfaces is also drier than last sampling day. The decrease rate of soil water content in lower layers is much less than that on the surface layer. For the layer at about 50-60cm depth, soil water content drops about 4% under biogenic crust and about 1.5% under sand surface from last sampling day. At about 30-40cm, the decrease is about 5.5% for biogenic crust and 2.5% for sand surface. Similar dropping also can be seen at the depths of about 10-20cm and 80cm. This indicates that the dropping rate under biogenic crust is much higher than under sand surface. The difference of dropping rate may be due to the difference of evaporation from the two surfaces and other soil water properties under the two surfaces.

Another sampling was conducted on February 16, 1998. The surface was obviously wet because there was another heavy raining process between February 10-15 and total rainfall is about 29.3mm. In some lower places can see muddy Playa and even some water on the ground. The vegetation began to bear bloom and some annuals were coming out, some even had flowers. A difference method was used for the sampling so that the result was more comparable. The top sand samples was taken by using a scraper to collect the top thin layer soil. The depth of scraping was about 0.5cm. The top biogenic crust sample was taken by the same method. A scraper was used to spade the top surface layer with a depth of about 0.5cm. The samples were also taken from soil profiles at various depths. The sampled soil was hold in plastic bags, which were sealed to prevent water lose before weighting. Enough amount of soil was taken for each sample so that it was not necessary to have more than one sample. The bag has about 3.5g in weight. The measurement result of this sampling date is also plotted in Figure 8.3b.

Two features are important in the sampling result on February 16. First the surface of sand seems to be drier than that of biogenic crust. Sand surface had volumetric water content of 8.62% and biogenic crust high up to 11.54%. The difference was about 3%. Second, the soil undergoes two processes of drying and wetting. Due to rain in about 2-3 days ago, the upper layers were under wetting process and the depth can be up to about 30cm under biogenic crust and 50cm under sand surface. However, the layers below this depth still undergo the drying process. In both soil profiles can be clearly seen these two processes. Compared to February 3, soil water content of surface layer and the layer at about 20cm depth increases obviously but the content of the layers at >50cm depth decrease at various amount. For soil profile under biogenic crust, the decrease at about 60cm reaches maximum.

8.3.4 Soil water content in late wet season

I continued my soil sampling on March 28 and the measurement result is shown in Figure 8.3c. There were only small rains since February 16 but the total precipitation was only 8mm. The surface of the field was dry by eye observation but soil still remained obviously wet at about 15-20cm depth under both sand and biogenic crust surfaces. In previous samplings, I took the soil samples from two different surface patterns under their natural condition. That was the sand surface on top dune and biogenic crust in interdune. Due to the difference of these two patterns in geomorphological and meteorological aspects, the results may not really be comparable. In order to minimize the impact of environmental factors on measurement results, soil samples were taken from the interdune with very close to each other. The soil samples under sand surface were taken from the patch where the original surface was biogenic crust and now removed and under biogenic crust from the place adjacent to the patch within about 10m. In order to have a good representative sample, the top sample was collected on the surface and the amount of the sample was enough for the measurement. Six layers were sampled under the two surfaces.

This sampling result validates what was gained in the last sampling. Three important features of soil water content can be seen in the result. Again, sand surface has lower soil water content than biogenic crust. However, the layers under sand surface have greater soil water content. At about 50cm, volumetric soil water content is about 8.96% under sand surface, which is about 0.5% greater than that under biogenic crust. The difference at 20cm is much more obvious. The soil water content is about 6.17% under sand surface and it is only 5.14% under biogenic crust, with a difference of up to 1%. Another feature is that at the depth of above 100cm from the surface, the soil water content is similar (about 6.0%-6.5%) under the two surfaces.

Another sampling was conducted on April 10, 1998. The place for the sampling was the same as that on March 28. There was little rain in the region since March 28. So the surface was very dry, but the soil beneath the surface still kept the same features as seen on March 28. When digging to about 15-20cm from both sand surface and biogenic crust, the soil was found to be obvious wet. At the layers below this depth, it was difficult to distinguish the difference of soil wetness by eye observation. Sampling was done from the layers with the same depth as that on March 28 and the result is also plotted in Figure 8.3c for comparison.

Compared with the result of March 28, similar features of soil water content can be easily seen from the result of April 10. Again, soil water content is higher under biogenic crust. Due to no rain, the surface layer of biogenic crust only has volumetric water content of about 1.71%, which is about 0.7% greater than sand surface. However, the content is greater in the layers under sand surface. At 20cm depth, water content under sand surface still remains at the level of about 5.25% while the content under biogenic crust drops to about 3.72%. These features are important for preparing initial soil water content of the modeling.

Drying process prevails in late wet season due to little rain. However, different drying rate can be seen for the soil profile under the two surfaces. At 20cm, soil water content drops from 6.17% of March 28 to 5.25% of April 10 under sand surface, with a decrease of 0.92%. The content drops from 5.14% to 3.71% under biogenic crust, with a decrease of 1.43%. The drop rate of the later is about 55% higher than the former. Drop rate at the depths of 50cm and 80cm is similar. This probably implies that soil moisture escapes more rapidly from biogenic crust in the long run.

8.3.5 Soil water content in extremely dry season

The change of soil water content in wet season is necessary for understanding the ecological and meteorological processes in the arid region, which may have profound impact on the surface temperature change of the region. However, we are more interested in its change in dry season because we aim at modeling surface temperature change in the season when the anomalous LST phenomenon for the study is observed to be more obvious. Two sampling activities were carried on in the dry hot season for measuring soil water content. They were on June 13 and July 16 respectively. Due to above two months of no rain, the region was extremely dry and the temperature is high. The sky was very clear and most shrubs were in dormancy. The same sampling method was applied to the two surfaces. The results are shown in Figure 8.3d.

Compared with Figure 8.3c, two features can at least be seen from Figure 8.3d. First, soil water content gradually increases with the depths in Figure 8.3d while in

Figure 8.3c the content in upper layers may be higher than in lower ones. Second, the first two layers have very close water content in both cases. This indicates that the first 20cm soil from the surface is in extremely dry condition. Very close water content can also be seen in the layers lower than 80cm. From June 13 to July 16, the soil continues its drying process but the process mainly concentrates on the layers between 20-80cm. Soil water content below 80cm seems to be identical.

8.3.6 Conclusion of the measuring soil water content for simulation

A series of sampling had been carried for measuring soil water content of profiles under sand surface and biogenic crust. The sampling period ranged from wet to dry seasons in 1998. These measurements produce an essential observation about the soil water change in the region.

After the heavy rain, soil water content is much higher in the top layer of biogenic crust. According to the sampling on January 25, which was about two weeks after heavy rain, top soil had about 18% volumetric water content for biogenic crust and 10% for sand. Water penetration in the soil is very slow. Within one week after heavy rain, soil below 40cm still remains very dry under sand surface (Figure 8.3a). After two weeks, the water finally penetrated into its maximal depths. Top soil experiences a fast drying process after rain. Within one week after heavy rain, the top 10cm soil under sand surface may lose its half water content (Figure 8.3a), due to both penetration and evaporation. However, the top layer also easily gets into wetting by rains. There were several small rains and a heavy rain between February 3 and 16. The rains made the top two layers under both surfaces much wetter on February 16 (Figure 8.3b). Soil profiles undergo different drying processes after the rain. It seems that the drying rate is faster under biogenic crust than sand surface. From March 28 to April 10 there is no rain and the volumetric water content of soil profile decreases 0.92%, 0.89% and 0.55% at depths of 20cm, 50cm and 80cm respectively under sand surface and 1.42%, 0.95% and 0.72% under biogenic crust surface (Figure 8.3c).

In dry season, the region is very dry. Soil water content in the surface layer is only about 0.5%-0.8%. One interesting phenomenon is that the upper layer has a little bit higher soil content under biogenic crust though the difference is very weak due to both extremely low. However, the soil profile under sand surface may have higher soil water content in lower layers. According to the sampling on July 16, 1988, the content is about 2.31% at 40cm under sand surface, which is about 0.5% greater than under biogenic crust. This difference is probably ascribed to different speed of the soil water attenuation in the two soil profiles.

It seems that seasonal raining only affects the soil water change of top 1m profile. Soil water content seems to be identical at the layers below 1m in wet season and below 0.8m in dry season. At this depth, the content is about 6.0%-6.2% in wet season (Figure 8.3b and 7.3c) and 5.0%-5.5% in dry season (Figure 8.3d). Similar results had also been reported in Yair et al. (1997).

Soil moisture data of the region needs to be carefully set for the modeling. From the above sampling results one can see that the surface is extremely dry in the hot summer and the soil profile loses its water content rapidly after the rains. Volumetric soil water content in both sand and biogenic crust is less than 1%. Under about 80-100cm, the soil water content may reach up to 5.0%-5.5% and variation below this depth is very small. Therefore, this content can be used as the lower boundary soil water content for the modeling. It should be note that the soil water content under two surface patterns is different. The content under sand surface is slightly higher. Usually the difference is about 0.5%-0.7% at about 40cm depth and about 1% at 60cm. This difference may have important impact on the observed LST contrast on both patterns (see chapter 4).

8.4 Determination of soil density, porosity and constitution

Soil density and porosity strongly affect many soil thermal properties related to the change of surface temperature. Therefore, they also need to be determined in the study.

8.4.1 Soil bulk density

Due to many difficulties, soil density can not be directly measured in the field. The determination of soil density is generally done through soil samples in laboratory (Campbell, 1985). For the study, I also follow this general method. The sampling was carried on July 16, 1998 when the surface was very dry. For sand surface, the soil in top 10-15cm is loose and below this it is compressed into solid granular structure. For biogenic crust, the surface is covered with dry crust and under the crust (about 1-5mm) is sand, which is granulated into solid structure. Thus, soil density can be measured by using soil samples taking from the solid layer of the profiles.

Four metal cylindrical cans were used to sample the soil for the measurement of density. The cylindrical can has a volume of about 460cm^3. Two samples were taken from profile under biogenic crust and two under sand surface at about 40cm and 100cm respectively. At these levels, the sandy soil is dry in both profiles by eye observation. Soil water content measurement in previous section indicates that both profiles have similar volumetric soil water content (about 5.5%) at about 100cm depth. However, at 40cm depth the profile under sand surface has about 4.2% of volumetric water content, which is about 0.5% greater than under biogenic crust. In order to compare their net density, the samples were firstly dried in an oven with 105°C for 24 hours and then weighted with an electric scale. The bulk density is calculated as the ratio of net weight to the volume of the can.

Table 8.3 Soil density under sand and biogenic crust surfaces

Samples	Depth (cm)	Net weight (g)	Density (g/cm^3)	Density (kg/m^3)	Average (kg/m^3)
Biogenic crust	40	689.2	1.498261	1498.261	1510.544
	100	700.5	1.522826	1522.826	
Sand	40	686.7	1.492826	1492.826	1502.00
	100	695.4	1.511174	1511.174	

The result shown in Table 8.3 indicates that soil bulk density of the two soil profiles is almost the same. Thus, it can be concluded that the soil bulk density of the region is about 1500 kg/m^3. However, small difference between the two depths can also be seen. The density at 100cm depth on both soil profiles is slightly higher than at 40cm. This is rationale because the deeper the soil is in, the more it is compressed by gravity. Another small difference is between the two profiles. The bulk density under biogenic crust is also slightly higher than under sand surface. One of the reasons probably is that the soil profile under biogenic crust is well protected which makes the soil more compressed than under the loose sand surface. However, this difference is negligibly small.

8.4.2 Soil porosity

Soil porosity is an important factor affecting heat transfer in soil. Many soil properties strongly related to both the volume and the shape of soil pore and the media filled in the pore (Campbell, 1985; Hausenbuiller, 1978). As described in chapter 6, soil porosity is needed to estimate soil thermal conductivity, heat capacity and other parameters under study. The same soil samples for soil density were used to determine soil porosity. It is well known that the soil in its natural status is composed of soil solid materials (mainly mineral and organic) and its pore, in which some space may be occupied by soil water (solution) and the rest is occupied by air and water vapor (Marshall and Holmes 1979). Thus, we have

$$V_t = V_s + V_p = V_s + V_w + V_a \tag{8.1}$$

where V_t is total soil volume, V_s is the volume of soil materials (sand in the region), V_p is volumetric space (pore) of the soil, which is equal to the space occupied by water (V_w) and air (include water vapor) (V_a). Thus, if the total soil volume V_t and volumetric space of the soil V_p or V_s are given, one can easily calculate the soil porosity as the ratio of V_p to V_t. However, directly measuring both V_s and V_p by geometric method is extremely difficult (Hillel, 1980).

It can assume that water has a density of about 1 g/cm^3 and the space in soil can be fully filled by water at saturation. Thus, using water to fill the space of soil sample up to saturation, one can measure the maximal water content of the sample and then calculate its V_p. However, natural soil usually contains some water and its vapor in the pore. Therefore, before using water to fill the soil pore, I firstly dry the samples in an oven for 24 hours so that the pore is only occupied by air. Usually measuring water volume is not as easy as its weight. A scale is used to weight the soil samples. Soil pore volume can be computed as $V_p = W_b - W_a$, where W_b and W_a are the weights of soil sample before and after filled with water.

One problem is that the experiment of filling water should be done carefully and slowly enough so that the filling process does not destroy the inner pore structure of the original soil body. I used a fabric to cover the soil sample surface in the filling process so that the water permeates slowly into the soil samples used for the experiment. The filling stopped when the soil sample is observed at saturation with water.

Table 8.4 Porosity of soil profiles under sand and biogenic crust surfaces

Samples	Depth (cm)	Weight W_b (g)	Weight W_a (g)	Pore (cm^3)	Porosity (%)	Average (%)
Biogenic crust	40	689.2	898.6	209.4	45.5217	45.2391
	100	700.5	907.3	206.8	44.9565	
Sand	40	686.7	900.1	213.4	46.3913	45.7718
	100	695.4	903.1	207.7	45.1522	

From the result listed in Table 8.4 one can see that the porosity of the two soil profiles is very closed. Thus, it can be concluded that the sandy soil porosity of the region is about 45%. Small difference can be seen between the two depths and the two profiles. In both profiles, soil porosity at lower layer is slightly less than at upper layer. Porosity at 40cm depth is about 0.5%-1.2% lower than at 100cm. This is because the soil in lower soil layer is more compressed than upper layer under effect of gravity. Thus, porosity in lower layer is less when other factors equally function. Another small difference is between the two profiles. Porosity of biogenic crust is slightly lower. This is also mainly because, as indicated in the section about measuring soil density, soil profile under biogenic crust surface is denser. However, this difference is too small to be considered in the modeling.

8.4.3 Soil particle density

There is a difference between bulk density and particle density. As indicated above, bulk density is the ratio of total solid mass to its total volume. The particle density, also termed soil density for short, is the mass divided by its particle volume. And the modeling requires the particle density of soil to compute its thermal properties. Thus, the particle density of the sandy soil has to be determined. Particle volume of soil sample depends on its porosity. Actually, total volume of soil is constituted of soil particle volume and its pore space. This gives the possibility of estimating soil particle density through porosity and bulk density. The derivation is as follows:

$$D_s=M_s/V_s=M_s/(V_t-V_p)=\{M_s/V_t\}/(1-V_p/V_t)=D_b/(1-P_s) \tag{8.2}$$

where M_s is soil mass, D_s and D_s are soil particle and bulk density respectively and P_s is soil porosity. Thus, soil particle density can be calculated as the bulk density divided by 1 minus porosity. Using this relationship, the particle density of the soil profile in the region is 2758.48kg/m^3 under biogenic crust and about 2769.68kg/m^3

under sand surface according to my sampling. This indicates that the sandy soil density is about 2750-2770kg/m^3. This conclusion is very close to the result of 2700kg/m^3 given by Marshall and Holmes (1979) after De Vries (1963) for quartz.

8.4.4 Soil constitution

Soil constitution of the region is very simple. It is the fine sand with various grain sizes (Tsoar 1990). The identical constitution is mainly attributed to the similar formation of the soil in the region: the aerial deposition of dust carried by strong wind (Gerson et al., 1985). Thus, the constitution of soil profile is very similar under the various surfaces. The only difference is the thin surface layer. Under bare sand, soil constitution is only sand. Biogenic crust is a mixture of silt, clay, microphytes and sand (Danin 1989). Thus, its constitution can be viewed as silt and clay with small fraction of fine sand.

The above qualitative analysis is not enough for the modeling, which needs a quantitative description of the soil properties. In fact, natural soil is a mixture of materials but not a pure one. Marshall and Holmes (1979) gave about 5% of silt and clay contained in sand. Campbell (1985) also estimated about 3%-5% of clay in sand. For the region, I will use about 3% of clay and silt in the lower layers of soil profile and the rest is sand. As to the top layer, a 100% of sand will be used for sand surface and 5% for the biogenic crust.

8.5 Air temperature and relative humidity

Both air temperature and relative humidity are required for the modeling as the basic inputs. The meteorological data of these two parameters available for the study was measured at about 2m above ground at Nizzana Research Site in July 1998.

The average air temperature change has been plotted in Figure 8.4. Though there are fluctuations in the curves, the general trend of the change can be clearly seen. No obvious differences of average air temperature can be found in each 10 days of July 1998. Average air temperature of the region reaches the level of high up to about 36°C at early afternoon (14:00-15:00) when the remote sensor AVHRR passes. Minimum air temperature occurs at about 5:00-6:00 when the temperature is in the range of 20-23°C (Figure 8.4). Thus, daily air temperature vibration of the region is up to above 13-15°C. Air temperature increases sharply in morning and decreases quickly in afternoon. After midnight, air temperature experiences a slow change. This represents the typical meteorological feature of arid environments. Because air temperature change is relative identical, the average of the whole month will be used in the modeling.

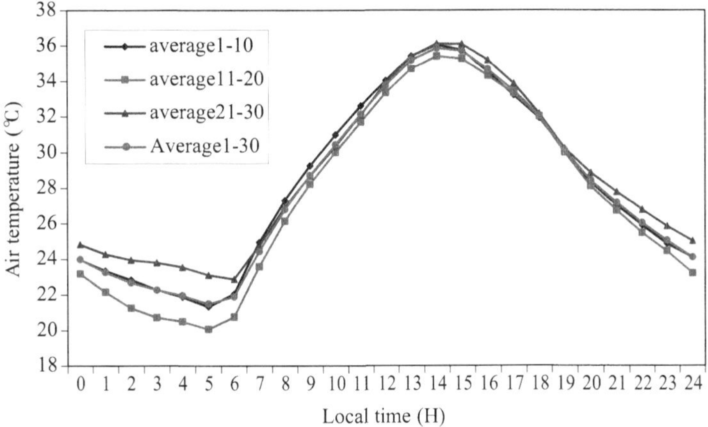

Figure 8.4 Air temperature of the region in July 1998.

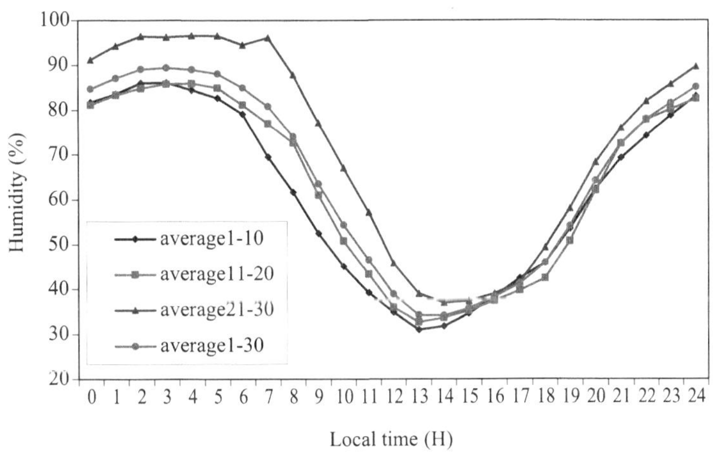

Figure 8.5 Air relative humidity of the region in July 1998.

The average change of air relative humidity is shown in Figure 8.5. Relative humidity not only relates to water vapor pressure but also depends on air temperature. When air temperature is high, relative humidity is low. This explains the change shape of relative humidity is opposite to the change of air temperature. During nighttime, air temperature is low but the relative humidity is high. Air temperature increases in morning during daytime, which leads to the decrease of relative humidity. Minimal air humidity occurs at about 14:00, when it is only about 30%-40%. Maximal humidity exists in 2:00-5:00, when it may reach up to 85%-95%. Another interesting feature shown in Figure 8.5 is that the difference of average air relative humidity is much more obvious in morning than in afternoon. The average July 1-10 has the lowest air relative humidity in 0:00-12:00 while average July 21-30 the highest. The maximal difference between the two extremes is up to 25%. From afternoon to midnight, average humidity is more identical. Based on these features, the average of the whole July was computed for the modeling.

8.6 Wind speed and roughness length

Wind speed is an important factor in micrometeorological modeling (Diak and Whipple, 1993; Dolman, 1993). The connection of wind speed to surface temperature change is that it directly affects the amount of sensible and latent heat fluxes which control the cooling process of the surface. The stronger the wind, the smaller the air resistance for sensible heat transferring into the air. Roughness length also has some relationship with wind speed though it is mainly depended on the morphological characteristics of the surface.

8.6.1 Wind speed

Two months of wind speed data are available for the analysis of its daily change: April and July 1998. The daily change of wind speed at Nizzana Research Site is clearly shown in Figure 8.6, from which several features can be seen. First of all, there is a peak in late afternoon when wind speed is observed to be maximal. The peak is between 15:00-16:00 in April (Figure 8.6a) and between 16:00 and 18:00 in July (Figure 8.6b). The value of wind speed during the hours is about 4m/s in April and 4.8m/s in July 1998. Second, wind speed in daytime is generally greater than in nighttime. After reaching the peak in late afternoon, wind speed quickly declines in evening. Figure 8.6 indicates the speed drops to the minimum (about 1m/s) at midnight. In summer (Figure 8.6b) the ascent of wind speed accelerates in late morning hours during the day till later afternoon at about 17:00. Third, the vibration of wind speed is greater in July than in April. Difference between daytime maximum and nighttime minimum is about 4.5m/s in July and 3.2m/s in April. Fourth, due to the spatial variation of geomorphological shapes a periodical cycle of regularity is clear in the daily wind speed change under general governance of local meteorological conditions. Another feature is that the daily change of wind speed tends to be more identical in July. In April, a relatively greater vibration can be found in the hours 0:00-14:00 (Figure 8.6a). The average wind speed in July (Figure 8.6c) will be used as the input of the model.

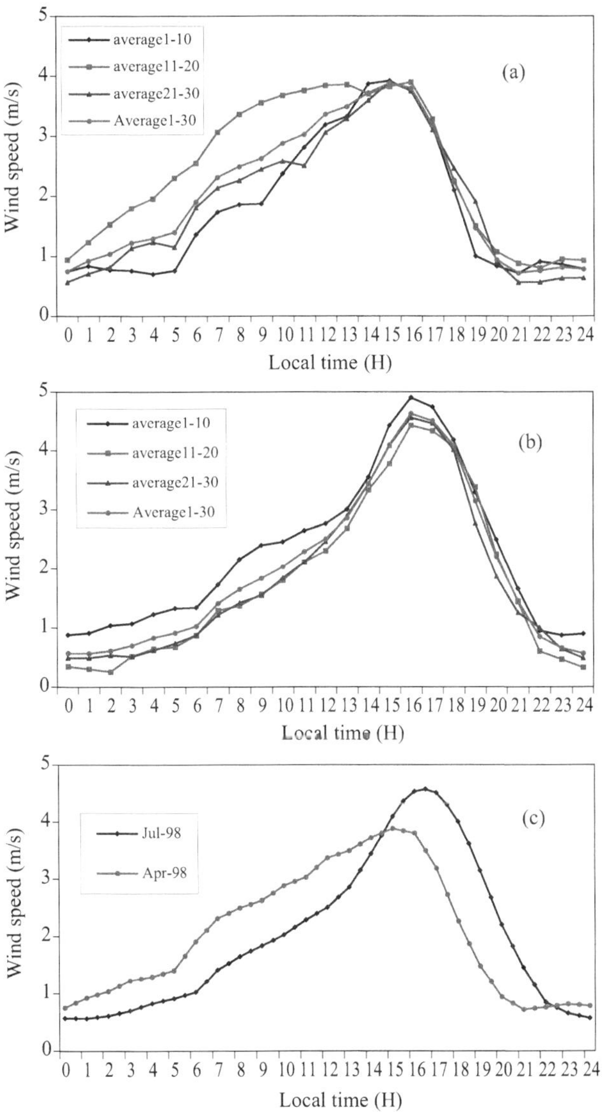

Figure 8.6 Wind speed of the region, showing the average change of every ten days in April 1998 (a) and July 1998 (b) and the whole month (c).

8.6.2 Roughness length

The determination of roughness length (z_0) is a process of complex derivation and calculation, which requires the supports of many profile data especially the sensible heat flux and wind speed (Brutsaert, 1982). The focus in the model, however, is on surface temperature and heat flux change, including sensible heat flux, which is unknown and needs to be computed for the estimation of surface temperature, For the sake of simplicity, I determine this parameter through literature.

Due to its importance in energy balance models for evaporation estimation, many studies have been oriented to the determination of roughness length under various situations (Brutsaert, 1982; Monteith, 1973). The amount of this parameter strongly depends on surface conditions. Specifically, it has a close correlation with the height (h) of vegetation. Choudhury et al. (1986) calculated this parameter for wheat as $z_0=0.13h$, where h is the height of wheat in his study. Szeicz et al. (1969) found that roughness length on fully developed maize and sorghum for wind speeds of approximately 2.5m/s was $0.105h$ and Garratt (1984) stated that $0.1h$ could be a rough estimation of z_0 though its range can vary from $0.02h$ to $0.2h$. Based on their data, Van Bavel and Hillel (1976) reported a roughness length of $z_0=0.01$m for bare soil in an open area. After reviewing the literature of relevant studies, Brutsaert (1982) gave a proportion of $z_0=h/8$ for a broad range of vegetation types.

The surface in the study region is scattered with some shrubs with a height less than 0.5 m. Therefore, according to the proportion given by Brutsaert (1982), we get $z_0=0.062$ m. Based on the rough estimation of Garratt (1984), we have $z_0=0.05$ m. Please note that this proportion is for the dense vegetation types. The study region is an arid desert and shrub only occupies about 17.5% of the surface on the Israeli side and about 4.5% on the Egyptian side. The situation of vegetation type in the region is far from that assumed for the proposed proportion of the parameter to the height. Thus, it can be concluded that the parameter in the region is between the bare soil ($z_0=0.01$m) and the dense vegetation ($z_0=0.05$m). And the value $z_0=0.02$m is used for the modeling.

9 Simulation of LST change in the region

The purpose of the simulation is to compare the difference of daily performance of surface temperature change and heat flux variation on the two most important patterns (sand and biogenic crust) and to distinguish the leading factors governing the difference. Analysis of the simulation results will finally lead to uncovering the mystery of LST anomaly in the region.

9.1 Modeling LST change of biogenic crust and sand

The simulation of surface temperature change is carried on under two data sets: one for biogenic crust and one for sand surface. Therefore, all required parameters have to be carefully designed according to the two surfaces. Fortunately, the soil under the two surfaces is similar due to the same formation process in the region. This provides an easy way of preparing soil property data for the modeling. The required data have been described in previous chapter.

According to the situations of the region and due to the extreme scarcity of knowledge about the micrometeorological change and soil physics of the Egyptian side, several assumptions have been constructed for the simulation. First of all, some meteorological indicators such as global radiation, air temperature, air humidity and wind speed near the surface are assumed to be little variation on both sides. This assumption allows us to focus our attention on the variation of sub-surface soil properties and surface features for understanding the LST anomaly on both sides.

Second, surface temperature change of the Israeli side can be represented by the thermal performance of biogenic crust due to its above 72% ground is covered with this pattern. The Egyptian side can also be represented by sand for the same reason in terms of its surface constituents. The surface in the real world is in fact a mixed one composed of several patterns, as indicated in chapter 6. Simulation through surface energy balance equation requires a specific data set for the surface under consideration. For a mixed surface, some soil parameters are really difficult to properly estimate through surface composition. Therefore, the way to analyze the LST change on both sides is through comparison of thermal variation of the two typical surface patterns.

Third, the lower boundary conditions of soil profile for solving the soil heat transfer and soil water movement equations are assumed to be the same for the two typical surface patterns. No daily change of soil temperature and soil water content is assumed at about 0.5m from the surface. Actually, daily heat penetration into the soil is within 30-50cm depth in arid environment (Hausenbuiller, 1978). According to the reports of Fania and Zipora (1997) and Zemel and Lomas (1977) who monitored soil temperature regime in Israel, the assumption of no daily soil temperature change at about 50cm depth is reasonable for the south Israeli desert environment.

9.2 Performance of daily surface temperature changes on the two surfaces

The most important output of the simulation is the LST change of the two surfaces under consideration. The simulated daily surface temperature change of sand and biogenic crust in dry summer is shown in Figure 9.1a, which also plots the real average air temperature change as a reference for comparison. Several features are explicitly illustrated in Figure 9.1a. First of all, surface temperature of biogenic crust and sand surface in summer can reach as high as about 48-50°C at about noon when the temperature peak occurs. This range is quite well corresponded to the observed change through ground truth measurements and on remote sensing images according to the spatial LST distribution on the Israeli side. Remember the data used for the simulation represents the general situation in summer and is measured at Nizzana Research Site, which is adjacent to the border (see Figure 5.4). Thus, the simulation result based on the data only reveals the general LST change of the two surfaces in the area adjacent to the border. This does not deny the possible higher or lower performance of the surface temperature peak on specific day under specific conditions such as in the area far from the border on the Israeli side, where very high LST can be found (see Figure 4.19 and 4.20).

Second, the biogenic crust has obviously higher surface temperature during daytime. The temperature difference between the two surfaces disappears during nighttime. As shown in Figure 9.1b, The maximal LST difference between the two surfaces is high up to 2.8°C at about noon. This implied that biogenic crust is much hotter under the beating of strong solar radiation in summer when little soil water is available for evaporation. However, the difference is not obvious during nighttime. This change of daytime and nighttime LST difference is extremely important in the explanation of thermal variation on both sides. The anomalous LST phenomenon was observed to be obvious during daytime at about 14:00 *in situ* satellite pass but this temperature difference is very obscure on nighttime image of remote sensing, as described in chapter 3. More important is that the simulated surface temperature at about midnight is about 22°C (Figure 9.1a) and the sharp LST contrast tends to disappear at this temperature level, as indicated in chapter 5.

Third, the surface temperature of biogenic crust and sand surface is much higher than air temperature during the day and is much lower than air temperature during the night. This is in accordance with the general situation of temperature change in arid en-

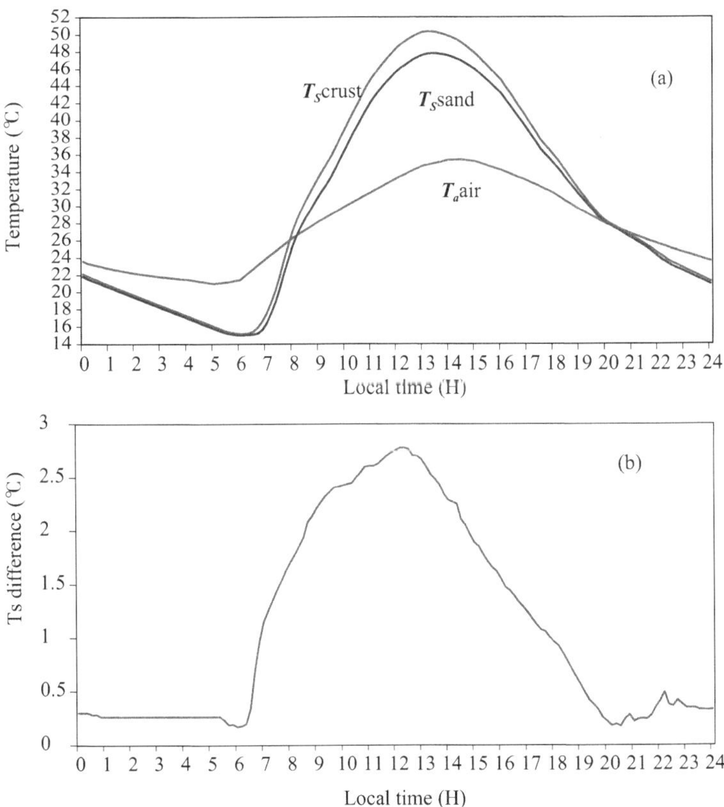

Figure 9.1 Comparison of daily surface temperature fluctuation on biogenic crust and sand surfaces, illustrating (a) LST change of the two surfaces and (b) LST difference between the two

vironment. Generally speaking, daytime of arid environment is very hot and nighttime is cool. Daily variation of air temperature is up to 15°C and of surface temperature is even higher (up to 35°C).

Fourth, the surface temperature increases rapidly in early morning and declines rapidly in late afternoon. The minimal surface temperature can be seen at about 5:00 when it is about 15°C. Surface temperature starts to ascend from 6:30 and its maximal ascendance can be seen in the hours from 7:00 to about 9:00. During this period it climbs rapidly from about 16°C to 33°C. In afternoon, the rapid descent of surface temperature occurs in the hours from 16:00 to about 19:00. It drops from about 42°C to 27°C. This change reveals the general heating and cooling process of the region in dry summer.

9.3 Sensitivity analysis of the model

The parameters for sensitivity analysis are the important ones governing surface temperature change of the region. Totally 22 parameters are selected to test the sensitivity of the model. These parameters are listed in Table 9.1.

What we concern with the modeling is the surface temperature change under the governance of these parameters. Thus, sensitivity analysis of the model is to test the change of surface temperature with the small change of these parameters. The output of the model for the data set of biogenic crust surface was used as the reference of the comparison for the sensitivity analysis. Because we are more interested in the change of surface temperature in early afternoon when satellite passes, the average surface temperature in 12:00-14:30 was used as the reference of surface temperature for the analysis. As to the small change of the parameters, we define as 1%. Thus, for a specific parameter, the analysis is to answer the following question: how many percent of surface temperature change would be resulted from the change of 1% in the parameter under consideration?

Table 9.1 Sensitivity analysis of the simulation model

Parameters	Change of T_s in ℃	Change of T_s in %
Roughness length	-0.04287	-0.09302
Surface emissivity	-0.02166	-0.04699
Surface albedo	-0.05914	-0.12831
Soil thermal conductivity: quartz	0.00138	0.00300
Clay minerals	0.00144	0.00313
Soil air	-0.00272	-0.00589
Soil composition: quartz	0.00294	0.00638
Clay minerals	0.00631	0.01371
Organic materials	-0.00213	-0.00462
Soil density	-0.00321	-0.00696
Soil specific heat	-0.00321	-0.00696
Initial soil water content	0.00557	0.01209
Lower layer soil water content	0.00818	0.01775
Initial soil temperature	0.00097	0.00212
Lower layer soil temperature	0.00299	0.00650
Air temperature	0.34519	0.74904
Air relative humidity	0.00017	0.00038
Wind speed	-0.05514	-0.11965
Global radiation	0.12829	0.27838
Soil layer thickness	-0.01062	-0.02303
Surface layer thickness	-0.00098	-0.00214
Saturated hydraulic conductivity	0.00625	0.01357

Table 9.1 indicates that a 1% increase in the parameters would let to the various change of average surface temperature T_s in 12:00-14:30. The largest change in T_s is observed in the small change of air temperature and global radiation. When air temperature increases 1%, average surface temperature will increases about 0.345°C, with 0.749%. An increase of 1% in global radiation will result in 0.128°C increase in surface temperature, with 0.278%. Actually air temperature and global radiation are the two essential factors of the model. Global radiation determines the amount of energy available for the micrometeorological and hydrological process in the soil-air interface system. And air temperature governs the exchange of energy between ground and the air. Their sensitivity to LST change in the model implies their importance in

the simulation.

A 1% increase in surface albedo, roughness length and surface emissivity corresponds respectively to a decrease of 0.059°C, 0.043°C and 0.022°C in surface temperature, which account for 0.128%, 0.093% and 0.047%. An increase of 1% in soil layer thickness would let to about 0.01°C decrease in surface temperature, accounting for 0.023%. The model appears to be insensitive to the change of other parameters, including soil thermal conductivity, soil composition structure, initial soil temperature and soil water content and so on. As indicated in Table 9.1, a 1% change of these parameters would let to less than 0.01°C or less than 0.02% change in surface temperature. Their insensitivity implies that the modeling result of LST change will not be obviously shaped by the possible small error of these parameters in preparation of the required data sets for the modeling.

9.4 Comparison of surface energy balance

Only a part of the global radiation received by the ground surface is absorbed hence enters the energy balance process of the land-air interface system. The remainder of the radiation is reflected back toward the atmosphere without affecting the surface temperature of the ground soil. The density of reflection or albedo largely determines the energy that drives the thermal movement of the ground soil. This can be clearly shown by the simulation results of the region. The comparison to the change of main parts of surface energy balance is shown in Figure 9.2. Due to much lower albedo of biogenic crust, the net radiation is much higher on biogenic crust than on sand surface during the day (Figure 9.2a). Average net radiation on biogenic crust is high up to 585.69 W/m^2 in the hours 12:00-15:00, which is about 26.42% higher than that on sand surface. This difference of net radiation is the main source of energy driving the thermal dynamics of the two surfaces.

The incident radiant energy that is not reflected by the ground surface is absorbed and converted into heat, which is then dissipated in several ways (Brutsaert, 1982). Under arid environment where little water is available for evaporation that consumes a lot of heat flux on the surface, sensible heat is the main way of energy dissipation of the ground surface. In the sand dune region, sensible heat into the air accounts for above 85% of net radiation in the hours 12:00-15:00. However, the amount of the heat on biogenic crust is much greater than that on sand. At about noon (12:00-13:00) when the surface temperature reaches its peak (Figure 9.2a), the average sensible heat into the atmosphere is about 550.13 W/m^2 on biogenic crust and 423.24 W/m^2 on sand surface. The former is above 1/4 (29.98%) greater. Because the sensible heat is proportional to the temperature difference between the surface and the air near the surface, the greater the heat means the greater the temperature difference under the same conditions of wind speed and surface roughness length. Given the same air temperature, this means that biogenic crust surface is much hotter.

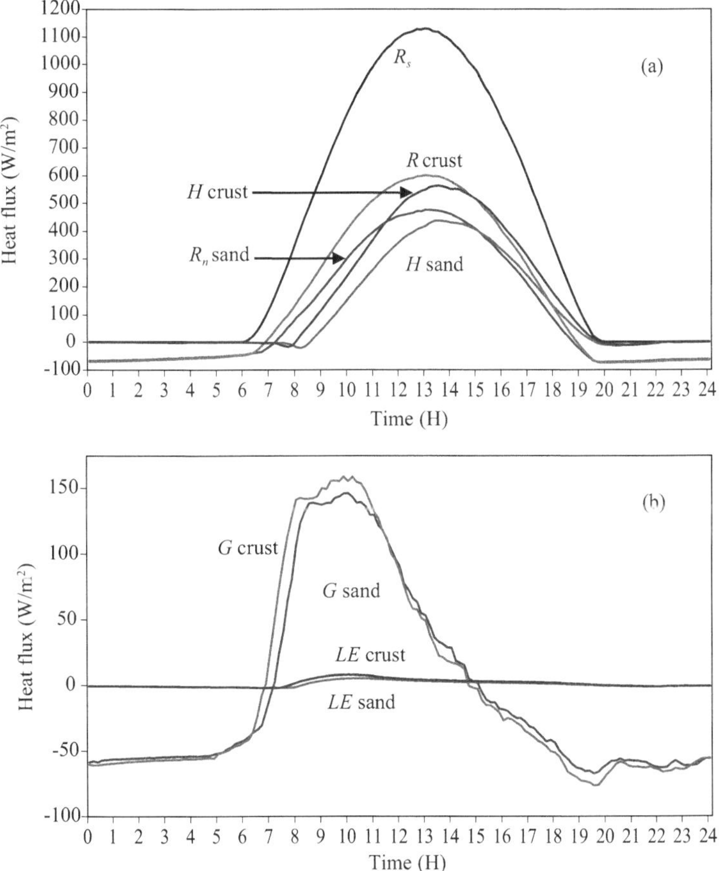

Figure 9.2 Comparison of simulated surface energy balance on biogenic crust and sand surfaces, illustrating (a) net radiation and sensible heat flux, and (b) soil heat and latent heat fluxes.

The change of soil heat and latent heat flux is shown in Figure 9.2b, from which one can see that the two heat fluxes are much less than sensible heat into the air. In the case of biogenic crust, average soil heat is only about 63.14 W/m^2 at noon 12:00-13:00, which only accounts for about 10.68% of the net radiation and 5.65% of the global radiation. And at this time, average latent heat is only about 4.36 W/m^2, accounting for 0.74% of the net radiation. Soil heat flux on sand is a little bit greater. At about noon, average soil heat on sand is about 68.12 W/m^2 and latent heat 3.36 W/m^2, accounting for 14.67% and 0.72% of its net radiation. This implies that heat transfers more strongly into the soil on sand surface, while the evaporation, in spite of very weak, is relatively stronger from biogenic crust at about noon. Compared with net radiation and sensible heat flux, conclusion can be drawn that the amount of these two parts is too small to have significant effect on the surface temperature change.

From the latent heat flux we can estimate the net lose of soil moisture per day in hot dry season. Total amount of latent heat flux is 270.15 W/m^2 on biogenic crust and 190.33 W/m^2 on sand surface. Thus, total evaporation is about 0.066kg/day from biogenic crust and 0.047kg/day from sand. The negative latent heat flux during nighttime also indicates that there is a process of dew formation, which releases some energy for the balance of the system. The total negative latent heat flux is about 77.95 W/m^2 on biogenic crust and 90.52 W/m^2 on sand, corresponding to 0.019kg and 0.022kg of dew formation respectively. This simulation result of dew formation amount is very close to the measurement of *Kidron* (1999) in the region. The main reason of more dew forming on sand surface during the night probably is that it has a relatively lower LST. Balanced from evaporation and dew absorption, net soil moisture lose in the arid environment is about 0.047kg/day on biogenic crust and 0.025kg/day on sand. The result also indicates that evaporation is a little bit stronger from biogenic crust. This explains the difference of soil moisture content observed in hot summer in the layers under the two surfaces.

The change of the above main parts of the surface energy balance experiences a different process in nighttime when there is no solar radiation. Figure 8.2a indicates that net radiation and sensible heat into the air are negative on both surfaces. And negative soil heat flux and latent heat flux can also be seen in Figure 9.2b. This implies that during nighttime the surface energy reaches its balance by getting heat flux from both air and soil. However, The heat flux released from both soil and air is quite small (< 80 W/m^2). Another feature is that the heat flux seems to be stable during nighttime. The dew forming on the two surfaces during nighttime quickly evaporates in early morning when the sun comes up. Thus, the maximal latent heat appears at about 9:00-10:00 (Figure 9.2b) and it is about 7 W/m^2 for biogenic crust and 5 W/m^2 for sand. This evaporation process explains the rapid ascendance of the surface temperature in morning.

9.5 Change of main parameters in the surface energy balance

Analysis of the change of main parameters in the system can also be assistance to understand the difference of surface energy balance on the two surface patterns.

As indicated above, sensible heat flux into the atmosphere is the leading way of dissipating the energy absorbed by the ground. According to equation (6.1.5), sensible heat into the air is proportional to the land-air temperature difference and inversely proportional to air heat transfer resistance. Given the air density and heat capacity be little variation in the temperature range, sensible heat into the air is mainly contributed by the land-air temperature difference. On the other hand, air heat resistance also plays an important role in governing sensible heat flux (Lhomme et al., 1994). Figure 9.3a compares the air heat resistance of the two surface patterns. Generally speaking, the change of air heat resistance change is similar for the two surface patterns, especially during daytime. This is mainly because air heat resistance is determined by wind speed, which is assumed to be the same for the two surface patterns. Though very close of air heat resistance between the two surfaces, slight difference can still be seen (Figure 9.3a), with greater on sand. During the day, land-air temperature difference is big and wind speed is high. Thus, heat is easily transferred into the air. From 8:30-19:30 air heat resistance of the two surfaces is rather small. Minimal air heat resistance occurs in late afternoon at about 16:00-17:00 when wind is the strongest. Maximal resistance occurs at about midnight when wind speed declines to the lowest. During nighttime air resistance is much higher. Very big air heat resistance during nighttime makes the sensible heat flux quite small.

Surface resistance to evaporation is also a very important factor in determining evaporation rate hence latent heat flux (Garratt, 1984; Van Bavel et al., 1976). In arid environment soil water content is extremely low, surface resistance to evaporation is

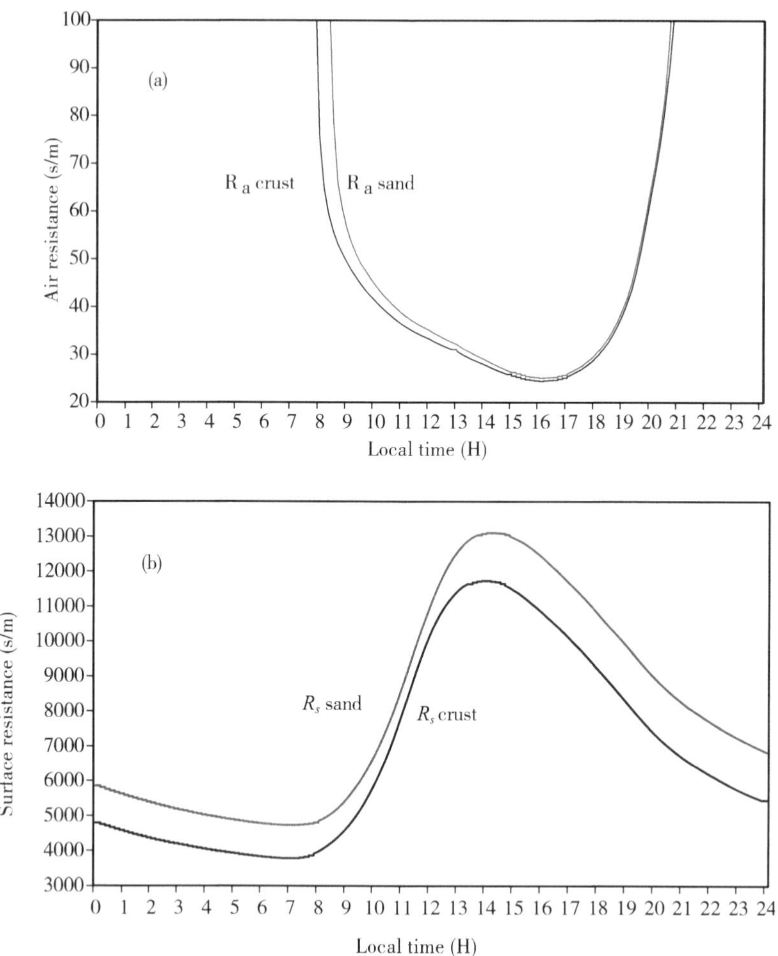

Figure 9.3 Comparison of air heat resistance (a) and surface resistance to evaporation (b) on biogenic crust and active sand surfaces.

very big. As indicated in Figure 9.3b, surface resistance to evaporation is generally greater on sand than on biogenic crust. This is because biogenic crust has a little bit higher surface soil moisture content (Figure 9.4a). Maximal resistance occurs in early afternoon when surface water content is the lowest due to consecutive evaporation in the previous hours. Surface resistance to evaporation is high up to 13000 s/m on sand and 12000 s/m on biogenic crust in early afternoon. Minimal surface resistance to evaporation can be seen in dawn when the surfaces have the highest moisture content due to their absorption of moisture from the air during nighttime. Actually the change of surface resistance to evaporation is in accordance with the fluctuation of soil moisture content in the surface layer.

Availability of soil moisture is one of the three critical conditions for evaporation. The small evaporation in the border region (Figure 9.2b) is mainly due to its extremely low level of soil water content. Figure 8.4a compares the change of volumetric water

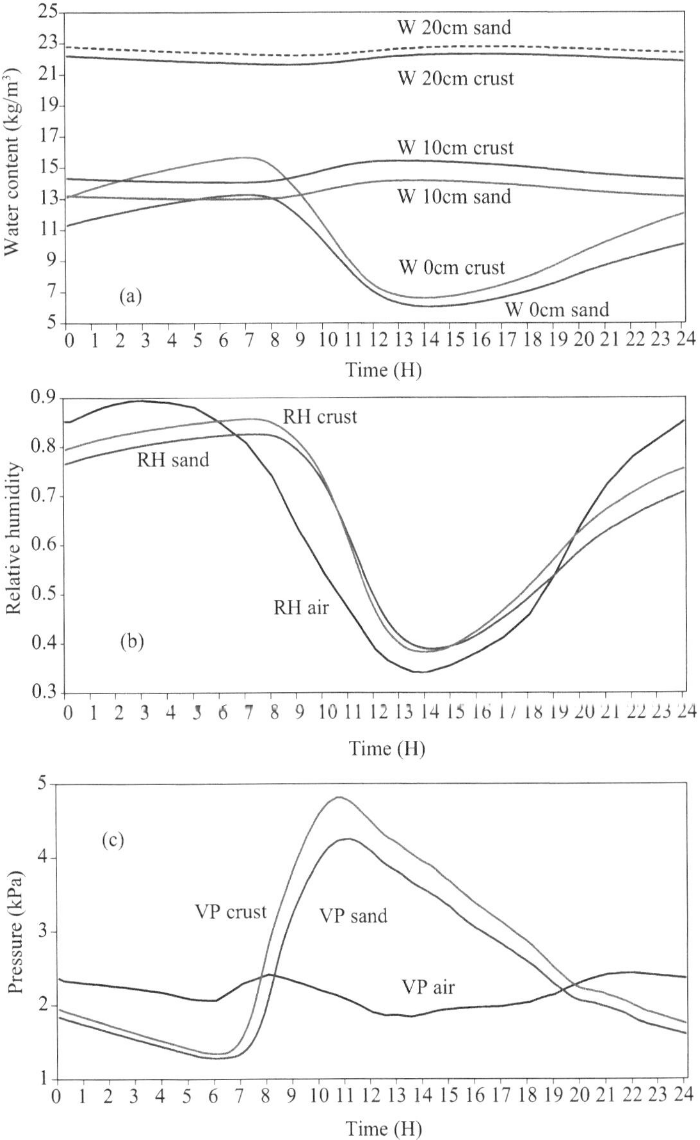

Figure 9.4 Comparison of (a) soil moisture content, (b) surface relative humidity and (c) surface vapor pressure on biogenic crust and sand surfaces.

content of the two surface patterns. The surface moisture change shown in Figure 9.4a is consistent with our measurements of soil water change given in previous chapter. Surface moisture of biogenic crust is obviously higher than that of sand, even though their absolute difference is only about 0.6kg/m^3 during the day. In morning surface moisture tends to decline because of evaporation. The moisture content reaches minimum in early afternoon at about 13:00-15:00 when LST is very high. In nighttime the surfaces absorb some moisture from the air, which makes their moisture content gradually increasing till dawn at about 6:00(Figure 9.4a).

Comparison of soil moisture content at 10cm and 20cm depth also is plotted in Figure 9.4a. Due to little heat flux penetrating into the lower layers of the profiles, soil moisture content becomes little fluctuation beneath 20cm depth. At 10cm depth, soil moisture has very small daily change and at 20cm the change is little. Another important feature is that soil moisture content at 10cm depth is greater under biogenic crust surface than sand surface. However, situation at 20cm is opposite. Soil moisture content under biogenic crust is less. This is consistent with what was observed in the measurements of soil water content, presented in previous chapter.

Under unsaturation, soil relative humidity is mainly depended on soil water content though temperature is also an important determinant. Figure 8.4b shows the change of soil relative humidity on the two surfaces. The shape of the change is similar with that of surface moisture content. During nighttime biogenic crust has a little bit higher surface relative humidity. In morning the relative humidity of sand surface is higher. However, in afternoon, biogenic crust returns to have slightly higher relative humidity. This maybe is because the soil under biogenic crust has higher hydraulic conductivity and the biogenic crust has greater evaporation rate (Figure 9.2b).

Comparison of surface vapor pressure on the two surfaces is illustrated in Figure 9.4c, which also plots air vapor pressure as a reference. According to equation (7.1.9), evaporation from the surface occurs when vapor pressure of the surface is greater than that of the air and dew formation happens when air vapor pressure is greater. From Figure 9.4c one can see that vapor pressure of the two surfaces is greater than air vapor pressure from about 8:00 to about 20:00. This means that evaporation from both surfaces happens during this period even though it is rather weak. Dew formation also occurs in the arid environment during nighttime because air vapor pressure is greater.

However, comparison of the area between surface and air vapor pressures reveals that evaporation is greater than vapor absorption, as indicated in the previous section. Figure 9.4c also indicates that biogenic crust has higher surface vapor pressure than sand surface. This implied that evaporation from biogenic crust is relatively stronger. The lower vapor pressure of sand surface during nighttime also indicates more dew forming on it, as indicated in previous section.

9.6 Contribution of the leading factors to surface temperature difference

Based on the above modeling results, I try to isolate the contribution of the leading factors to the LST difference on the two surfaces. Assumed the same effects in terms of such meteorological factors as global radiation, wind speed and air temperature, the leading factors of the model that may lead to the LST difference can be distinguished as albedo, surface emissivity and sub-surface soil properties.

The difference of surface albedo between biogenic crust and sand plays a leading role in making the LST difference on the two surface patterns. As indicated in previous chapter, biogenic crust has albedo of about 13% lower than sand. Thus, under the same global radiation, biogenic crust absorbs much more energy from the incident radiation. Simulation indicates that the difference of albedo accounts for above 86% of net radiation difference between the two surfaces. As shown in Figure 9.2b, the average difference of surface temperature between the two patterns is 2.63°C in the hours 12:00 15:00. Using the albedo of biogenic crust and the soil properties under sand surface to simulate produces an average surface temperature of 49.25°C in the same period, which is about 2.31°C higher than LST of sand and 0.32°C lower than LST of biogenic crust. Therefore, it can be concluded that about 87.83% of LST difference between the two surfaces are contributed by their albedo difference.

The simulation indicates that surface emissivity difference has little effect on the LST difference. Using the emissivity of biogenic crust and soil properties of sand to simulate, the contribution of surface emissivity difference to the LST difference has also been tested. The result indicates that the surface emissivity difference can only lead to about 0.05°C of LST change from the LST of biogenic crust. This implies that the difference of surface emissivity can only explain about 1.90% of the LST difference between the two surfaces.

The difference of soil properties also has important effect on the LST difference. As mentioned above, soil properties are different under the two surfaces. The soil under biogenic crust has higher water content in the upper layers and lower soil moisture content in the lower layers. Consequently, evaporation from biogenic crust surface is greater. Due to the different soil constituent, the surface layer of sand has much higher thermal conductivity. Higher thermal conductivity of sand makes its heat transfer a little bit faster into the sub-surface soil. Moreover, higher heat capacity in the sub-surface soil under sand surface due to its higher soil moisture content also makes it more difficult to be heated. Soil heating process has proportional correlation to the change of surface temperature and evaporation. Under the combined functioning of the two aspects, the surface soil of biogenic crust is a little bit easier to be heated. And this finally shapes the different change of temperature regime and energy balance on the two surfaces.

It is really difficult to separate the effect of every soil properties on LST change because they are interrelated and mutually depended. Therefore, I roughly count their effect as the remainder of the total minus the effect of albedo and ground emissivity. By this ways, the difference of soil properties accounts for about 10.27% of the total LST difference between the two surface patterns.

Given the strong solar radiation in summer, the thin surface layer (1-5mm) of biogenic crust can be easily heated in spite of its relatively higher moisture content. Under the arid environment, soil water content is extremely low and evaporation can only consume a very small part of total energy absorbed by the surface. As indicated in Figure 9.2b, in 12:00-15:00 evaporation only accounts for about 0.61% of net radiation in the case of sand and for 0.65% in the case of biogenic crust. This small percentage of latent heat flux plays little effect on the cooling down of ground surface. As a result, the surfaces experience different heating processes due to their difference of absorbing incident sky radiation. Obviously, the surface energy absorption accumulates more in the upper biogenic crust layer than the sand surface layer. And this makes the surface of biogenic crust hotter than sand surface. Even though the involved process is rather complicated, the mechanism is very clear, *i.e.* the greater amount of main surface flux components on biogenic crust makes its surface temperature higher to reach its surface energy balance at higher level.

9.7 Understanding thermal variations on both sides of the region

This above analysis and explanation of surface temperature difference on the two typical surface patterns can be extended into the understanding of surface temperature difference and thermal variation on both sides. The Israeli side has about 72% of biogenic crust cover and the Egyptian side above 80% of sand surface. Their surface albedo difference is also very obviously, as indicated in previous chapter. The lower surface albedo on the Israeli side makes it absorb much more incident solar energy and atmospheric emittance. Though most of the net radiation returns into the atmosphere as sensible heat, the remainder is still obviously greater on the Israeli side and this remainder is much strongly related to the surface and soil temperature change and soil moisture movement. During summer, the region is very dry. Most vegetation is in dormancy (Karnieli and Tsoar, 1995; Otterman and Tucker, 1985). Leaves of shrubs reduce to minimum and more than two third of canopy is dead. The dead canopy is not only out of functioning evaporation but also absorbs more incident radiation due to its darker tone hence lower albedo. Even though the Israeli side has more vegetation, the evaporation from both the ground and the dormant shrubs is still very small in comparison with the net radiation. Together with the transpiration contribution of vegetation, the average evapotranspiration on the Israeli side is estimated to be about 41 W/m^2 in 12:00-15:00, accounting for 7% of the net radiation. This small latent heat flux has little effect on the surface energy balance process in the arid environment. LST change in the arid environment is then mainly controlled by the amount of incident solar energy absorbed by the ground as soil heat. Therefore, the anomalous LST change and thermal variation on both sides of the region can still be explained as the direct result of their obvious albedo difference, as Otterman and Tucker (1985) interpreted. This albedo difference is mainly caused by the sharp contrast of surface composition on both sides. As mentioned above, the Israeli side has much more biogenic crust while the Egyptian side is mainly dominant with bare sand. Considered this sharp contrast, the LST anomaly observed on remote sensing images can be understood as the result of spectral functioning difference of biogenic crust and sand on both sides of the region.

Soil properties on the Israeli side can be viewed as similar with those of biogenic crust and on the Egyptian side similar with those of sand. This analogy can be used to understand soil heat transfer on both sides of the region. As mentioned in above section, there are more heats accumulating on the surface of biogenic crust than on sand, due to the difference of their soil properties especially thermal conductivity. This functioning process also exists on both sides of the region because of their surface composition difference. Most surface on the Israeli side is covered with biogenic crust which has lower thermal conductivity while the surface of the Egyptian side is dominates with sand having relatively higher thermal conductivity. Accordingly, one can expect that the Israeli side has more heat accumulation on its surface due to its relatively lower thermal conductivity. Finally, the ground of the Israeli side has higher surface temperature. Therefore, except the main contribution of spectral functioning difference of biogenic crust and sand, sub-surface soil properties as a whole are the second main factor leading to the anomalous LST change and thermal variation observed on both sides of the region.

9.8 Comparison with Charney's classic studies and other related studies

In the mid-1970s, Charney and his co-workers conducted a series of studies about the interaction between desertification and climate in Sahel and other dry-land regions (Charney, 1975; Charney et al., 1975; 1977). They proposed a biogeographysical feedback model, also dubbed albedo model (Williams and Balling, 1996). On the basis of their investigation in Sahel and Sahara drought, Charney and his co-workers argued that the overgrazing due to settlement and over-population in arid and semiarid regions will lead to the reduction of vegetation cover hence the increase in surface albedo. Consequently, the increased albedo will decrease the surface net radiation, result in surface cooling and lower ground surface temperature, promote subsidence of air aloft, reduce convection and cloud formation, therefore reduce regional rainfall in the arid and semiarid regions. Additional drying in the Sahel region leads to regional climatic desertification which positively feeds back to the overgrazing and follows the circling procedures. Using a general circulation model developed for meso-scale regional atmospheric and climate simulation, they demonstrated the important effect of albedo in regional climate especially rainfall in Sahel, Sahara and other dry-land regions. The prepositions and assumptions underpinning the studies become known as Charney hypothesis (Williams and Balling, 1996). Basing on the simulation with two different surface albedos (low albedo 14% represents low human stress and high albedo 35% the high stress due to overgrazing), Charney et al. (1975) found that the mean rainfall over Sahara during the calendar month of July decreases about 43% from low-albedo to high-albedo experiments. And the results from the high albedo experiment is quite close to the observed distribution of rain in Sahara summer. Thus, they concluded that surface albedo can have a substantial effect on climate in Sahara and very local changes in albedo may be sufficient to produce droughts.

It seems that the higher surface albedo leading to higher surface temperature in the border region of our study supports Charney hypothesis about the function of albedo in climate change. Close comparison still finds some differences between our study and Charney's ones. In Charney's studies, it is assumed that the overgrazing leads to the reduction of vegetation cover, which in turn leads to the increases in surface albedo. Thus, the importance of albedo in Charney's biogeophysical feedback model is based on the direct relationship between vegetation and surface albedo. In the border region, the obvious difference of surface albedo across the border is not only due to the vegetation difference, but also due to the biogenic crust difference (Karnieli & Tsoar, 1995). Overgrazing on the Egyptian side not only causes the reduction of vegetation, but also prevents the establelishment of biogenic crust on the surface. Actually it is the biogenic crust difference that leads to the substantial difference in albedo across the border. Desert plants, which still have much higher evapotranspiration, contribute more less to the albedo difference across the border. Despite the support from a variety of modeling studies, the Charney hypothesis was vigorously challenged on theoretical and empirical grounds (Williams and Balling, 1996). Since Charney's model is based on the direct contribution of vegetation to albedo, its assumption about the higher surface albedo leading to the higher surface temperature has not been seen in other arid regions. This is mainly because the lower albedo contributed by more vegetation cover still have higher evapotranspiration that prevents the increase of surface temperature. On the other hand, since the bare surface as a result of vegetation removal is easy to be dried hence has very low top-soil moisture that prevents soil heat transfer into lower layers, the surface with high albedo contributed by low vegetation still has relatively higher surface temperature. This has been observed across the US-Mexico border (Balling, 1979). Only in the region where lower surface albedo is not dominantly contributed by vegetation and where soil moisture is very low such as our region, the phenomenon of higher surface albedo leading to lower surface temperature is possible.

Another important condition for the occurrence of surface temperature anomaly across border is the very low soil moisture in the region. Many numerical modeling studies showed that the soil moisture is also very important in affecting climate change in the arid and semiarid regions (Williams and Balling, 1996). Cunning & Rowntree (1986) demonstrated that Sahara climate is not only related to surface albedo change but also to soil moisture changes and the importance of soil moisture could override the changes to the albedo. Combining the local energy balance model to a global climate model, Lanicci et al. (1987) found that soil moisture changes in the dry-lands of the American Great Plains could have an important impact on the location and intensity of convective storms. Franchito and Rao (1992) developed a global model that includes detailed biogeophysical feedbacks to simulate the effects of desertification and found that desertification causes a reduction in local net radiation, soil moisture, evaporation and precipitation which consequently results in an increase in local surface and near-surface temperature. In our region, the soil moisture is extremely low during the hot dry summer. Our field sampling found insignificant difference in soil moisture of the top layer under bare sand and biogenic crust. The insignificant difference in soil moisture between the two main surface patterns which respectively represents the surface on both sides makes it easy to understand the relatively less importance than albedo in leading to the occurrence of the surface temperature anomaly across the border.

9.9 Summary of the chapter

An anomalous thermal phenomenon was observed in the daytime remote sensing images in the sand dunes across the Israel-Egypt border. This interesting phenomenon has not been thoroughly studied for understanding of its occurrence, especially from the viewpoint of micrometeorological modeling. A complete surface energy balance model that couples soil temperature change simultaneously with soil moisture dynamics has been establelished in the paper in order to simulate the surface temperature change and heat flux variation of the region. Various measurements have been conducted at the Nizzana Research site on the Israeli side of the region for the required simulation data. Using the data and the numerical solution methodology of the model, the surface temperature and heat flux change has been simulated on the two most important surface patterns of the region: biogenic crust representing the Israeli side and bare sand representing the Egyptian side.

The simulation results indicate that surface temperature of biogenic crust and bare sand in summer can reach as high as about 48-50°C. According to the model, biogenic crust does have a surface temperature of about 2.5-2.8°C higher than bare sand in the hours 12:00-15:00. During the night, difference of surface temperature between biogenic crust and bare sand is not obvious. Sensitive analysis of the model indicates that the largest change in T_s is observed in a small change of air temperature and global radiation, which is followed by surface albedo, roughness length and surface emissivity. A 1% increase of air temperature and global radiation will lead to 0.75% and 0.28% increase of LST. And the same amount of surface albedo, roughness length and emissivity increase will result in 0.13%, 0.09% and 0.05% decrease of surface temperature. Comparison of surface energy balance reveals that average net radiation of biogenic crust can reach high up to 585.69 W/m^2 in the hours 12:00-15:00, which is about 26.42% higher than that of active sand. Above 85% of net radiation return into the air as sensible heat flux. Average soil heat flux of biogenic crust is about 63.14 W/m^2 at the noon hour 12:00-13:00, accounting for 10.68% of net radiation. The amount of soil heat flux is a little bit greater on sand surface than on biogenic crust. Due to little soil moisture available for evaporation, latent heat flux of biogenic crust is only 4.14 W/m^2 at noon, accounting for 0.74% of net radiation. For the bare sand, latent heat is only 3.36 W/m^2 and accounts for 0.72% of its net radiation. Daily evaporation of the region is very small under the arid environment. Net soil water loss due to evaporation is estimated to be about 0.047 kg/day on biogenic crust and about 0.025 kg/day on bare sand. Negative latent heat flux during the night also indicates the formation of some dews on both surfaces. But bare sand surface, due to its relatively lower LST at night, seems to obtain more moisture from the air than biogenic crust. This difference in surface energy balance also produces different change of subsoil properties especially soil moisture content at various depths and the surface vapor pressure and relative humidity. The mutual connection of these changes finally determines LST difference and thermal variation on the two typical surface patterns of the region.

The LST anomaly on both sides of the border region can be understood in several ways. First of all, surface composition difference between the two sides is the direct reason of the LST anomaly. The Israeli side is mainly composed of biogenic crust, which covers about 72% of the surface. Bare sand only accounts for about 7% of its surface. However, the Egyptian side is mainly covered with bare sand, which accounts for above 80% of the total area. Biogenic crust and vegetation only account for about 12% and 4.5% of the total surface area. With the assumption of similar atmosphere and similar subsoil composition and formation on both sides, this difference of surface composition directly explains the occurrence of the LST anomaly on both sides even though the functioning mechanics of the process is rather complicated.

Second, the surface composition difference on both sides leads to the difference of surface and subsoil properties. The combined functioning of this surface and subsoil property difference makes the LST change on both sides appear to be of difference in its expression. The coherent mechanics can be understood through the simulation results of surface energy balance model. According to their surface composition, the Israeli side can be represented with biogenic crust and the Egyptian side with bare sand. The simulation results indicate that biogenic crust and bare sand do have obvious LST difference under the assumption of the same global radiation, wind speed, air temperature and relative humidity. The LST difference between the two surface patterns can reach up to about 2.5-2.8°C in early afternoon when satellite passed and observed the LST of the region. The most important factor leading to the LST difference of the two surface patterns is surface albedo, which accounts for about 87.83% of LST difference between the two surfaces. Subsoil properties also have significant effect on the LST difference of the two surface patterns. It can explain about 10.27% of the LST difference. Though surface emissivity has significant difference on the two typical surface patterns hence on both sides, it can only account for about 1.90% of the LST difference between the two surfaces.

Third, the mechanism of the LST difference on the two typical surface patterns can be summarized as follows. The biogenic crust surface has much lower surface albedo. Thus, it absorbs much more incident sky radiation than bare sand surface. Even though most of the absorbed energy is dissipated as sensible heat into the air, the remainder of the absorbed energy is also much more on the biogenic crust than on the bare sand. This remainder energy is dissipated mainly in two ways: heat into the soil and evaporation. Because soil water content is extremely low in the arid environment, evaporation only accounts for about 0.5%-1% of the total net radiation. Therefore, the slightly higher evaporation on biogenic crust is not enough to cool down the much higher surface temperature due to its more absorption of the incident radiation in the arid environment.

The difference of soil composition in the surface layer plays important role in shaping LST change on the two surfaces. Due to more quartz content, thermal conductivity of surface layer on bare sand is greater than on biogenic crust even though the later has a little bit more soil moisture content. In the depth of about 10-15cm onward, the subsoil under bare sand has more moisture content than the subsoil under biogenic crust. Consequently, this changes the soil properties under the two typical surfaces. Higher soil thermal conductivity of surface layer on bare sand makes its soil heat transfer a little bit easier than it is on biogenic crust. In another word, heat is more difficult to go down into the lower layers under biogenic crust and easier to accumulate in the upper thin layer of biogenic crust. This makes its surface temperature even easier to ascend.

Finally, this mechanism of heating process difference on the two typical surface patterns can be extended to understand the LST anomaly on both sides of the region. During the hot summer season, the region is very dry and vegetation is in dormancy. Leaves of most shrubs reduce to minimum. Even though the Israeli side has more vegetation, the evapotranspiration contributed by the vegetation is still very small in comparison with the net radiation. This small latent heat flux has little effect on the surface energy balance process in the arid environment. LST change in the desert region is mainly controlled by the amount of incident solar energy absorbed by the ground as soil heat. Therefore, the anomalous LST change and thermal variation on both sides of the region can still be explained as the direct result of the obvious albedo difference on both sides. This albedo difference is mainly caused by the sharp contrast of surface composition on both sides as the result of spectral functioning difference between biogenic crust and bare sand. On the other hand, the Israel side has stronger evaporation rate than the Egyptian side. This means that the soil water content may be easier to lose on the Israeli side than on the Egyptian side where bare sand prevails.

After a long time of consecutive evaporation to the hot summer, the subsoil of Israeli side on average may have less soil moisture than that of the Egyptian side. And this property together with the low thermal conductivity of the crust may finally exacerbate soil heat accumulation on its surface layer (the thin biogenic crust). As a result, this makes its surface temperature easily increased. On the Egyptian side, relatively higher subsoil moisture content combined with the relatively greater thermal conductivity of the sand will speed up its soil heat transfer from surface into lower layers. This can also intensify the existing LST anomaly on both sides.

10 Conclusion

10.1 Problem of the study and its analysis

In the sand dune region across the Israeli-Egyptian border, an interesting phenomenon of thermal variation was observed on remote sensing imagery. The Israeli side with more vegetation cover has higher surface temperature than the Egyptian side where bare sand prevails. Moreover, this interesting phenomenon has not been deeply studied though it has been mentioned in some studies (Otterman et al., 1975; Otterman and Tucker, 1985; Qin et al., 2001a). The study intends to examine the phenomenon using the methodology of combining remote sensing analysis with micrometeorological modeling. The objectives of the study is to analyze the seasonal and spatial change of LST on both sides from the remote sensing data of NOAA-AVHRR and Landsat TM and to model the LST change and thermal variation from viewpoint of surface energy balance. The key factors controlling the change can be revealed through this examination. Methodology has been developed for LST retrieval from remote sensing data. A micrometeorological model and its numerical solution for simulating the LST change have been establelished. Besides, ground truth measurements have also been conducted for validation and simulation.

Remote sensing of LST is possible through detecting the thermal radiance from the ground (Qin et al., 1999). However, the observed radiance is not only from the ground but also from the sky. Besides, the thermal radiance emitted from the ground is attenuated when it travels through the atmosphere to reach the remote sensor. Water vapor of the atmosphere is the main absorber of the thermal radiance. Other gases such as CO_2 and air particles also play an important role in attenuating the thermal radiance on its way to the sensor in space.

On the other hand, LST change and thermal variation at the interface of land-air system is governed by a number of environmental factors. The important factors functioning in the system include incident sky radiation, wind speed, surface albedo, ground emissivity, soil properties, soil water content, etc. All of these factors have significant effect on the LST change and thermal variation on the ground surface. Actually, surface temperature change is a direct indication of the dynamic energy balance on the surface.

Considered the spatial adjacency, the atmosphere is assumed to be little variation on both sides. Consequently, the focus of the study is given to the difference of surfaces and its sub-surface soil properties for understanding the LST change and thermal variation on both sides.

10.2 Remote sensing of LST from NOAA-AVHRR data

Retrieval of LST from NOAA-AVHRR data is an important methodology in remote sensing. Several split window algorithms have been proposed in last two decades (Qin et al., 2001b). A complete derivation of the algorithm is generally omitted in the existing presentations. Moreover, some used many unnecessary assumptions to make the algorithms simple (Price, 1984; Becker and Li, 1990) while some involved the atmospheric parameters (Sobrino and Caselles, 1991; Prata, 1993) that are difficult to estimate due to unavailable profile data in most cases. These had led to many inconveniences for understanding the methodology and its application to the real world. Based on the thermal radiance transfer equation, a better split window algorithm involving minimal parameters (only transmittance and emissivity) is developed in the study with a full description of its mathematical derivation (Qin et al., 2001b). Along with this, the principle and method for the linearization of Planck's radiance equation, the derivation process of the algorithm and the method for determining the atmospheric transmittance are discussed with details. Emphasis is also given to sensitivity analysis of the algorithm for evaluation of probable LST estimation error due to the possible errors in transmittance and emissivity. Results from sensitivity analysis indicate that the algorithm developed in this study can provide a quite accurate estimation of LST from AVHRR data. The combination error of LST estimation due to both possible transmittance and emissivity is about 1.1°C, which is less than the generally accepted error 1.5°C (Kerr, 1992; Vogt, 1996).

10.3 Seasonal LST change in the region

NOAA-AVHRR data is used to analyze the seasonal LST change on both sides of the region. Using the split window algorithm developed in the study and the methodology for determining its coefficients, AVHRR data in 1995-1998 was processed for analysis of LST change on both sides (Qin et al., 2002a). The images with clear sky and viewing angle above 60° were selected for the processing. The average LST change of the border region was plotted in Figure 3.10-3.13, from which we can see the seasonal LST change on both sides of the region in early afternoon when AVHRR passed. In dry summer, LST of the region may reach up to 50-56°C while in winter it is about 28-35°C. Another feature shown in Figure 3.10 is that the Israeli side is generally about 2.5-3.5°C hotter in summer. In wet season, LST on the Israeli side is also higher (about 0.5-1.5°C). Only in a few extremely cases when the surface was very wet after heavy rain the Egyptian side had a little bit higher LST. However, the LST difference in these extreme cases is generally within -0.5°C.

The nighttime images of 1998 were also processed for LST retrieval. The result indicates that the LST change of the region at about midnight is mainly within the range of 20-26°C during summer and 10-15°C in winter. The obvious LST contrast on both sides disappears during nighttime of summer (Qin et al., 2002a). Sometimes a slightly higher LST can be seen on the Israeli side while sometimes on the Egyptian side. When temperature is low such as in winter, the Egyptian side will be a little bit hotter but the LST difference is within -0.75°C.

10.4 Retrieval of LST from Landsat TM images

Remote sensing of LST from the thermal band of Landsat TM data is generally an ignored area in comparison with the extensive application of its visible and near infrared (NIR) bands. Though brightness temperature can be computed from the digital number (DN) of Landsat TM thermal data using the equation provided by NASA, the algorithm for retrieving LST from the only one thermal channel still remains unavailable. Based on thermal radiance transfer equation, a mono-window algorithm has been developed in the study for retrieving LST from the only one thermal channel of Landsat TM data (Qin et al., 2001c). Three parameters are required for the algorithm: emissivity, transmittance and mean atmospheric temperature. Method about determination of atmospheric transmittance is given in the study through the simulation of atmospheric conditions with LOWTRAN 7 program. A practicable approach of estimating effective mean atmospheric temperature from local meteorological observation is also developed when atmospheric profile data is unavailable *in situ* satellite pass, which is generally the case for most studies such as this study. Sensitivity analysis of the algorithm indicates that the possible error of ground emissivity, which is difficult to estimate, has insignificant impact on the probable LST estimation error δT, which is sensible to the possible error of transmittance $\delta\tau_6$ and mean atmospheric temperature δT_a. However, average δT is generally within 1.5°C, which can meet the accuracy requirement of the study.

10.5 Spatial LST variation in the region

The mono-window algorithm has been applied to retrieve LST from available Landsat TM6 data of the border region. Figure 4.19 shows the spatial variation of LST distribution in March 29, 1995 at about 9:30 am. A clear difference of LST can be seen on both sides. The Israeli side has obviously higher LST. On average, LST is 33.87°C on the Israeli side and 31.01°C on the Egyptian side. Thus, the difference is about 2.86°C. High LST can be seen on the Israeli side in the right upper corner of the image. The LST in this area is high up to 36-38°C. Low LST concentrates in the middle part of the Egyptian side. The LST in this part is only about 28-29°C. In the area adjacent to the border is the LST at the moderate level, vibrating at about 31-32°C on the Israeli side and 29-30°C on the Egyptian side. A sharp contrast of LST still can be identified in this area adjacent to the border on both sides even though the difference is lower than the average one. This contrast of LST distribution on both sides makes the border clearly seen in the sand dune region.

10.6 Determining surface emissivity

Surface emissivity is a very important parameter in remote sensing of LST. Using soil samples taken from the field, experiments for determining ground emissivity have been done (Qin et al., 2005). The samples were put in an isothermal device (water bath) for measuring their radiant temperature. The emissivity of the samples can be calculated as the ratio of their radiant temperature to the temperature of the black plate reference. Actually the black plate is not a true black body and an emissivity of 0.999 was assumed for calibration.

Conclusion from the experiments indicates that the average emissivity of biogenic crust and sand is separately 0.97 and 0.95 in dry condition (Qin et al., 2005). Both biogenic crust and sand samples behave a weak increase of emissivity with temperature. And the increase pace of sand seems to be slightly greater. Comparison of emissivity of the samples in dry and saturation indicates that the difference of emissivity between biogenic crust and sand samples in dry condition seems to disappear at saturation when soil water governs the thermal properties of the soil. At saturation, both biogenic crust and sand samples have high emissivity of up to 0.983-0.984 for the treatment at 25°C and 0.985-0.987 at 45°C.

According to the surface structure of the region, average emissivity of the region is estimated to be about 0.968 on the Israeli side and 0.954 on the Egyptian side (Qin et al., 2005). Considered the fact that the Egyptian side has much more sand than the Israeli side, the estimation of its lower average ground emissivity seems to be reasonable.

10.7 Validation of LST change on the Israeli side

It is also important to know the surface temperature of main surface patterns and its change in different seasons. Using a handhold IR thermometer, several measurements of surface temperature on the typical surface patterns were conducted in various seasons. Though temperature is a changeable phenomenon, the measurements based on a number of sampling sites in a relatively short time still can reveal some trends of its variation. The emissivity of the main surface patterns was then used to correct RST into KST.

The results from ground truth measurements confirm what was expected according to the phenomenon observed in remote sensing images (Qin et al., 2005). In dry summer at about noon, the difference of surface temperature between the two surface patterns is high up to above 3°C. The difference is also obvious between morning and afternoon. In early and late dry seasons, the difference is still high up to above 2.5 °C at about noon. The difference disappears only within several days of heavy rain when the ground soil is very wet or at a level of high water content. This is because soil water content strongly shapes soil thermal properties governing the surface temperature change, as indicated in the chapter 7. However, the difference recovers in about two weeks after strong rain (Figure 5.5) though it is weak. Therefore, in wet season, the difference still appears in most of the days except the raining days and a few days after the raining process.

The ground truth measurements can also provide the information about the possible levels that the surface temperature of the main surface patterns may reach in various seasons. In dry summer season, surface temperature of biogenic crust may reach up to 52-55 °C at noon. The temperature of sand and playa may be up to 50-53°C and 51-53°C, respectively. Shrub canopy also has high temperature of about 35-38°C at noon in dry summer.

10.8 Estimation of surface composition in the region

The observed LST anomaly on both sides of the region is in fact the direct result of their structural difference of surface composition. The obvious difference of surface structure has been mentioned in various studies (Otterman, 1974; 1977; 1981; Karnieli and Tsoar, 1995; Tsoar and Karnieli, 1996) but up to now no quantitative data about the structure have been reported. The surface of the region can be classified into four categories: vegetation (mainly shrubs in dry season), biogenic (mainly cyanobacteria) crust, sand and playa (physical crust). Three methods have been employed to measure surface composition structure on both sides. Field observation and statistics has been applied to the Israeli side for measuring vegetation cover rate and the percentage of main surface patterns. Results from the field statistical analysis indicate that, on average, vegetation covers about 15.63% of the ground at Nizzana Research Site, biogenic crust occupies about 69.42%, sand 11.94% and playa 3.01%. These results are based on the statistical analysis of the five sampling plots in the field and the calibration according to a cross section involving two dunes and two interdunes.

The aerial photograph used for the measurement indicates that a sharp contrast of vegetation cover exists between the two sides (Qin et al., 2006a; 2003). According to the measurement, vegetation covers 17.02% of total area on the Israeli side and 5.81% on the Egyptian side. Biogenic crust occupies about 70.58% on the Israeli side and 18.03% on the Egyptian side. Sand covers about 8.89% on the Israeli side and 72.45% on the Egyptian side. Playa only covers a small percentage on both sides: 3.51% on the Israeli side and 3.72% on the Egyptian side. Biogenic crust and vegetation covers on the Egyptian side vanish rapidly with distance from the border. And the photograph only covers about 0.8km width of the Egyptian side. Therefore, lower vegetation cover rate and biogenic crust percentage, hence higher sand cover rate, are expected in inner Egyptian side.

The measurement results based on Landsat TM image strongly supports the results from the aerial photograph. Using the characteristics of NDVI on full vegetation surface and bare soil background, vegetation cover rate can be estimated from Landsat TM image. Measurement based on this principle indicates that vegetation covers 17.54% on the Israeli subset and 5.15% on the Egyptian subset. Distance from border also means different vegetation cover rate. On the Egyptian side, vegetation cover rate decrease with the distance from the border while the situation on the Israeli side is totally opposite.

Based on the three methods of measurement, the surface composition structure of the region is estimated as follow. On average, the vegetation cover rate is about 17.5% on the Israeli side and 4.5% on the Egyptian side. Biogenic crust occupies about 72% on the Israeli side and 12% on the Egyptian side. Sand covers about 7% on the Israeli side and 80% on the Egyptian side. And playa accounts for 3.5% on both sides (Qin et al., 2006a).

10.9 Impact of surface composition on LST change on both sides

The average LST change on both sides was computed according to the possible LST change of the four typical surface patterns. The result confirms what was gained in remote sensing analysis of LST change on both sides retrieved from the AVHRR and Landsat TM data. As shown in Figure 5.6b and 5.6c, when temperature is low such as in winter, the Israeli side will be cooler. Obvious LST contrast on both sides can be observed when temperature is at moderate to high level (>35°C). The difference tends to be minimal in 17-27°C, which occurs at around midnight of summer. This explains why a sharp LST contrast on both sides was observed in daytime images acquired at around 14:00 (AVHRR) and 9:30 (Landsat TM) but not in nighttime images acquired at around midnight. The estimation method of average LST change on both sides was applied to the ground truth measurements taken on the Israeli side. As shown in Figure 5.7, the average LST on the Israeli side is obviously higher for all the cases except those in winter when temperature is low.

10.10 Surface energy balance model and its numerical solution

A micrometeorological model has been establelished in terms of surface energy balance for simulating the surface temperature change and heat flux variation of the region (Qin et al., 2002b). The purpose of this modeling is to understand the mechanism leading to the occurrence of the LST anomaly and to reveal the key factors controlling the occurrence. Though various energy balance models have been proposed and used for various purposes, a complete model with full description of its complex factor relationship and its numerical solution is generally not available. The model establelished in the study couples soil temperature change simultaneously with soil moisture movement. The methodology for the numerical solution of the model has also been developed. Characteristics of the model can be summarized as follows. Soil heat and latent heat fluxes are determined by both soil temperature change and soil moisture movement, which can be described as differential equations. Crank-Nicolson implicit method is used to expand the differential equations into two sets of simultaneous linear equations, which are then solved by applying Gauss's elimination method. Latent heat flux is determined at the equilibrium when evaporation of the surface is equal to the soil water loss. And surface temperature is estimated as the heat fluxes of the surface reaching the status of balance. The iterative computation of Newton-Raphson method is used to approximate latent heat flux and surface temperature from the balances. Based on this complexity of the model's relationships, a detailed computation procedure is proposed.

10.11 Preparing the data for simulation

Simulation of LST change through surface energy balance model requires several data inputs especially global radiation, surface albedo, air temperature and so on. For preparation of these inputs, several measurements have been conducted at Nizzana Research Site (see Figure 7.4). These measurements include global radiation, surface albedo, air temperature and relative humidity, wind speed, soil water content, air temperature, soil density, soil porosity and soil constitution.

Global radiation was measured by pyranometers in summer (July16-20, 1998). Measurement results indicate that global radiation may reach up to about 1125 W/m^2 at about noon in a summer clear sky. The radiation gradually increases in morning and decreases in afternoon, just like a normal distribution curve (Figure 8.1). One-month data of air temperature and air relative humidity of the region is available for the study. The data was acquired in July 1998 which represents the general situation in hot dry season when the anomalous LST was observed to be very obvious on both sides. Average air temperature may reach up to 36°C in early afternoon (14:00-15:00). Minimal air temperature occurs at about 5:00-6:00 when it is about 20-23°C (Figure 8.4). Daily air temperature vibration is up to above 13-15°C. Relative humidity not only relates to water vapor pressure but also depends on air temperature. During daytime, air relative humidity is very low but it is high in nighttime. Maximal relative humidity occurs in the hours 2:00-5:00 when it may be up to 85%-95% and minimal one exists in early afternoon around 14:00 when it is only about 30%-40% (Figure 8.5).

Soil density and porosity are also important parameters for the modeling. According to our determination, soil bulk density has no significant difference in the profiles under biogenic crust and sand surface at about 40-100cm depth. Numerically, it is about 1500kg/m^3. For the particle density, it is about 2758kg/m^3, which is close to 2700kg/m^3 for quartz (Marshall and Holmes, 1979). Soil porosity also has no significant difference between the profiles under the two surfaces. For biogenic crust, it is about 45.23% and for sand 45.77%.

Two-month wind speed data are available for analysis of its daily change: April and July 1998. As shown in Figure 7.6 there is a peak of wind speed in late afternoon. In July it is about 4.8m/s and in April it is about 4.0m/s. Wind speed declines to minimal at about midnight when it is about 0.8m/s. During 23:00-14:00 average wind speed in April seems to be greater than in July. The opposite situation can be seen during 14:00-13:00 when it is greater in July than in April.

10.12 Determination of surface albedo

According to our hypothesis, the surface albedo is expected to be the key factor leading to the LST anomaly on both sides. Thus, careful determination of the albedo of typical surface patterns is extremely important for the study. Because albedo is defined as the ratio of reflected radiation to the incident one, the determination of surface albedo is done through the application of pyranometers to simultaneously measure the two components of radiation. The measurement period was the same as for global radiation.

As indicated in Figure 8.2a, sand has the highest albedo during the day, then followed by playa and biogenic crust. Vegetation has the lowest surface albedo. At about noon, the albedo of sand is about 46%. Biogenic crust has an albedo of about 13% lower than sand. Vegetation has an albedo of only about 20% at about noon and playa about 43%. According to their surface composition, it is estimated that the Israeli side has an average albedo of about 32% at noon and the Egyptian side about 44%, with a difference of about 12%. This difference is believed to be important in understanding the occurrence of LST anomaly on both sides of the region.

10.13 Measurement of soil water content at different depths

Surface temperature is strongly impacted by heat transfer in soil, and soil water content is a key factor shaping the transfer. Thus, knowing soil water content and its change is also very important in the study. Yair et al. (1997) monitored soil water content of the sand dunes in the rainfall years 1991-1993 for analysis of soil water percolation and movement in region. However, the data about soil water content of the region in recent years are not available. The comparison of soil water content under sand and biogenic crust surfaces also still remains undone. In order to have the most recent data about soil water content of the region, I took the measurements from August 1997 on the Israeli side. Due to the lack of proper instruments, classical gravimetric method was used for the measurements. In order to compare soil water content difference in the profiles under two typical surfaces, sampling spots were carefully designed to be in adjacency in the interdune.

Soil water content depends on both season and the time of sampling after rain. In dry season, the surface soil water content is very low. Volumetric content of soil water in the top layer of sand is only about 0.4%-0.5% and in the thin biogenic crust about 0.5%-0.7%. In wet season the content is also not very high. In about two weeks after the rain, the content drops to about 1%. Moreover, soil water content of sand surface is always less than that of biogenic crust. However, the soil profile at about 15-20cm depth onward has slightly higher water content under sand surface than biogenic crust. It seems that this is rational because sand surface has bigger pore that may prevents water transfer from lower level while the situation in biogenic crust surface is opposite. This difference of soil water content distribution at various depths may have significant effect on the LST difference between the two surfaces. At about 1m depth from the surface, volumetric water content of soil profile under the two surface patterns tends to be identical, which is about 5.0%-5.5%. Yair et al. (1997) also reported similar soil moisture content.

10.14 Analysis of the simulation results

The simulation results indicate that surface temperature of biogenic crust and sand in summer can reach as high as about 48-50°C. Biogenic crust do have a surface temperature of about 2.5-2.8°C higher than sand in the hours 12:00-15:00. During the night, surface temperature difference between biogenic crust and sand is not obvious. Sensitive analysis of the model indicates that the largest change in T_s is observed in a small change of air temperature and global radiation, which is followed by surface albedo, roughness length and surface emissivity. A 1% increase in air temperature and global radiation will lead to 0.75% and 0.28% increase in LST. And the same amount of surface albedo, roughness length and emissivity increase will result in 0.13%, 0.09% and 0.05% decrease in surface temperature. Comparison of surface energy balance reveals that average net radiation of biogenic crust can reach high up to 585.69W/m^2 in 12:00-15:00, which is about 26.42% higher than that of sand (Qin et al., 2002c). Above 85% of net radiation return into the air as sensible heat flux. Average soil heat flux of biogenic crust is about 63.14W/m^2 in 12:00-13:00, accounting for 10.68% of net radiation. The amount of soil heat flux is a little bit greater on sand surface than on biogenic crust. Due to little soil moisture available for evaporation, latent heat flux of biogenic crust is only 4.36W/m^2 at noon, accounting for 0.74% of net radiation. For the sand, latent heat is only 3.36W/m^2 and accounts for 0.72% of its net radiation. Daily evaporation of the region is very small under the arid environment. Net soil water lose due to evaporation is estimated to be about 0.047kg/

day on biogenic crust and about 0.025kg/day on sand. Negative latent heat flux during the night also indicates the formation of some dews on both surfaces. But sand surface, due to its relatively lower LST at night, seems to obtain more moisture from the air. This difference in surface energy balance also produces different change of sub-surface soil properties especially soil moisture content at various depths and the surface vapor pressure and relative humidity. The mutual connection of these changes finally determines LST difference and thermal variation on the two typical surface patterns of the region.

10.15 Understanding the LST anomaly phenomenon

The LST anomaly on both sides of the border region can be understood in several ways (Qin et al., 2002c). First of all, the surface composition difference between the two sides is the direct reason of the LST anomaly. The Israeli side is mainly composed of biogenic crust, which covers about 72% of the surface. Sand only accounts for about 7% of its surface. However, the Egyptian side is mainly covered with sand, which accounts for above 80% of its ground. Biogenic crust and vegetation only account for about 12% and 4.5%. With the assumption of similar atmosphere and similar sub-surface soil composition and formation on both sides, this difference of surface composition directly explains the occurrence of LST anomaly on both sides even though the functioning mechanics of the process is rather complicated.

Second, the surface composition difference on both sides leads to the difference of surface and sub-surface soil properties. The combined functioning of this surface and sub-surface soil property difference makes the LST change on both sides different. The coherent mechanics can be understood through the simulation results of surface energy balance model. According to surface composition, the Israeli side can be represented with biogenic crust and the Egyptian side with sand. Simulation results indicate that biogenic crust and sand do have obvious LST difference. The most important factor leading to the LST difference of the two surface patterns is surface albedo, which accounts for about 87.83% of LST difference between the two surfaces. Sub-surface soil properties also have significant effect on the LST difference, explaining about 10.27%. Though surface emissivity has significant difference on the two surfaces hence on both sides, it can only account for about 1.90% of the LST difference.

Third, the mechanism of LST difference on the two surfaces can be summarized as follows. Biogenic crust has much lower surface albedo. Thus it absorbs much more incident sky radiation. Even though most of the absorbed energy is dissipated as sensible heat into the air, the remainder of the absorbed energy is also much more on biogenic crust. This remainder energy is dissipated mainly in two ways: heat into the soil and evaporation. Because soil water content is extremely low in the arid environment, evaporation only accounts for about 0.5%-1% of the total net radiation. Therefore, the slightly higher evaporation on biogenic crust cannot cool down the high LST in the arid environment.

The difference of soil composition in the surface layer plays important role in shaping LST change on the two surfaces. Due to more quartz content, thermal conductivity of surface layer on sand is greater though biogenic crust has a little bit higher soil moisture content. In the depth of about 10-15cm onward, the sub-surface soil under sand has more moisture content. Consequently, this changes the soil properties under the two surfaces. Higher soil thermal conductivity of surface layer on sand surface makes its soil heat transfer a little bit easier. In other words, heat is more difficult to go down into the lower layers under biogenic crust and easier to accumulate in the upper thin layer of biogenic crust. This makes its surface temperature easier to ascend.

Finally, this mechanism of heating process difference on the two surfaces can be extended to understand the LST anomaly on both sides of the region. During summer, the region is very dry and vegetation is in dormancy. Leaves of most shrubs reduce to minimum and more than two third of canopy is dry or dead. The dead canopy is not only out of functioning transpiration but also absorbs more incident radiation due to its darker tone hence lower albedo. Even though the Israeli side has more vegetation, the evapotranspiration from both vegetation and ground is still very small in comparison with the net radiation. Considered the contribution of desert plants, the average evaporation in 12:00-15:00 on the Israeli side is estimated to be about 41 W/m^2, accounting for 7% of its net radiation. This small latent heat flux has little effect on the surface energy balance process in the arid environment. LST change in the desert region is mainly controlled by the amount of incident solar energy absorbed by the ground as soil heat. Therefore, the anomalous LST change and thermal variation on both sides of the region can still be explained as the direct result of their obvious albedo difference.

This albedo difference is mainly caused by the sharp contrast of surface composition on both sides. As mentioned above, the Israeli side has much more biogenic crust while the Egyptian side is mainly dominant with bare sand. Considered this sharp contrast, the LST observed on remote sensing images can be understood as the result of spectral functioning difference of biogenic crust and sand on both sides. On the other hand, the Israel side has stronger evaporation rate than the Egyptian side. This means that the soil water content on the Israeli side may be easier to lose than on the Egyptian side where sand prevails. After a long time of consecutive evaporation to summer, the sub-surface soil of the Israeli side in average may have less soil moisture. And this property together with relatively lower thermal conductivity of biogenic crust may finally accelerate soil heat accumulation on its surface (the thin biogenic crust). As a result, this makes its surface temperature easily ascended. On the Egyptian side, relatively higher sub-surface soil moisture content combined with relatively greater thermal conductivity of sand will speed up its soil heat transfer from surface into lower layers. This can also intensify the existing LST anomaly on both sides.

References

Asrar G, 1989. *Theory and Applications of Optical Remote Sensing*, John Wiley & Sons Ltd., New York, USA.

Badenas C, Caselles V, 1992. A simple technique for estimating surface temperature by means of a thermal infrared radiometer. *International Journal of Remote Sensing*, 13(15): 2 951-2 956.

Bahre C J, Bradbury D E, 1978. Vegetation change along the Arizona-Sonora boundary. *Annals of the Association of American Geographers*, 68:145-165.

Balling R C Jr, 1989. The impact of summer rainfall on the temperature gradient along the United States-Mexico border. *Journal of Applied Climatology*, 28: 304-308.

Balling R C Jr, Klopatek J M, Hildebrandt M L, et al., 1998. Impacts of land degradation on historical temperature records from the Sonoran desert. *Climatic Change*, 40: 669-681.

Barducci A, Pippi I, 1996. Temperature and emissivity retrieval from remotely sensed images using the "grey body emissivity" method. *IEEE Transactions on Geoscience and Remote Sensing*, 34(3): 681-695.

Barrett E C, Curtis L F, 1978. *Introduction to Environmental Remote Sensing*. John Wiley & Sons Ltd., New York, USA.

Becker F, 1987. The impact of spectral emissivity on the measurement of land surface temperature from a satellite. *International Journal of Remote Sensing*, 8(10): 1 509-1 522.

Becker F, Li Z L, 1990. Towards a local split window method over land surface. *International Journal of Remote Sensing*, 3: 369-393.

Ben Asher J, Mathhias A D, Warrick A W, 1983. Assessment of evaporation from bare soil by infrared thermometry. *Soil Science Society of America Journal*, 47: 185-191.

Ben-Asher J, Meek D W, Hutmacher R B, et al., 1989. Commutation approach to assess actual transpiration from aerodynamic and canopy resistance. *Agronomy Journal*, 81: 776-782.

Ben-Yosef N, Wilner K, 1985. Temporal behavior of thermal images. *Applied Optics*, 24: 284-286.

Ben-Yosef N, Wilner K, Fuchs I, et al., 1985. Natural terrain infrared radiance statistics: daily variations. *Applied Optics*, 24: 4 167-4 171.

Ben-Yosef N, Wilner K, Abitbol M, 1988. Modeling of natural terrain in the infrared. *Optical*

Engineering, 27: 928-932.

Berliner P R, 1988. *Contribution of Heat Dissipated at the Soil Surface to the Energy Balance of a Citrus Orchard*, PhD thesis, Hebrew University, Jerusalem, Israel.

Blad B L, 1983. Energy exchange among leaf, canopy and environment. *Crop-Water Relations*, John Wiley and Sons, New York, USA.

Bras L R, 1990. *Hydrology: An introduction to hydrologic sciences*, Addison-Wesley Publishing Company, Reading, Massachusetts, USA.

Braud I, 1998. Spatial variability of surface properties and estimation of surface fluxes of a savannah. *Agricultural and Forestry Meteorology*, 89: 15-44.

Bristow K L, Campbell G S, Papendick R I, et al., 1986. Simulation of heat and moisture transfer through a surface residue-soil system. *Agricultural and Forest Meteorology*, 36:193-214.

Brutsaert W, 1982. *Evaporation into the Atmosphere: theory, history and application.* Reidel D Publishing Company, Dordrecht, Holland.

Choudhury G J, Reginato R J, Idso S B, 1986. An analysis of infrared temperature observations over wheat and calculation of latent heat flux. *Agricultural and Forest Meteorology*, 37: 75-88.

Bryant N A, Johnson L F, Brazel A J, et al., 1990. Measuring the effect of overgrazing in the Sonoran Desert. *Climatic Change*, 17: 243-264.

Calvet J C, Jullien J P, 1996. Land surface Temperature Retrieval in Dry Conditions from infrared and microwave satellite radiometry. *Remote Sensing Reviews*, 13: 235-255.

Campbell G S, 1985. *Soil Physics with Basic: Transport models for soil-plant systems*. Elsevier Science Publishers, Amsterdam, The Netherlands.

Caselles V, Coll C, Valor E, 1997. Land surface emissivity and temperature determination in the whole HAPES-Sahel area from AVHRR data. *International Journal of Remote Sensing*, 18: 1 009-1 027.

Chen D X, Coughenour M B, 1994. GEMTM: a general model for energy and mass transfer of land surfaces and its application at the FIFE sites. *Agricultural and Forest Meteorology*, 68: 145-171.

Childs E C, 1969. *An Introduction to the Physical Basis of Soil Water Phenomena*, John Wiley & Sons Ltd., New York, USA.

Choudhury B J, 1989. Estimating evaporation and carbon assimilation using infrared temperature data: vistas in modeling. *Theory and Applications of Optical Remote Sensing*. John Wiley & Sons, New York.

Choudhury G J, Reginato R J, Idso S B, 1986. An analysis of infrared temperature observations over wheat and calculation of latent heat flux. *Agricultural and Forest Meteorology*, 37: 75-88.

Choudhury J B, Dorman J T, Hsu A Y, 1995. Modeled and observed relations between the AVHRR split window temperature difference and atmospheric perceptible water over land surfaces. *Remote Sensing of Environment*, 51(2): 281-290.

Coll C, Caselles V, 1994. Analysis of the atmospheric and emissivity influence on the split-window equation for sea temperature. *International Journal of Remote Sensing*, 15(9): 1 915-1 932.

Coll C, Caselles V, Schmugge T J, 1994. Estimation of land surface emissivity differences in the split-window channels of AVHRR. *Remote Sensing of Environment*, 48: 127-134.

Coll C, Caselles V, Sobrino A, et al., 1994. On the atmospheric dependence of the split-window equation for land surface temperature. *International Journal of Remote Sensing*, 15: 105-122.

Collatz L, 1966. *Functional Analysis and Numerical Mathematics*, Academic Press, New York.

Cooper D I, Asrar G, 1989. Evaluating atmospheric correction models for retrieving surface temperatures from the AVHRR over a tall-grass prairies. *Remote Sensing of Environment*, 27(1): 93-102.

Courault D Lagouarde J P, Aloui B, 1996. Evaporation for maritime catchment combining a meteorological model with vegetation information and airborne surface temperatures. *Agricultural and Forestry Meteorology*, 82: 93-117.

Cracknell A P, 1997. *The Advanced Very high Resolution Radiometer (AVHRR)*, Taylor & Fancis, London.

Cracknell A P, Hayes L W, 1993. *Introduction to Remote Sensing*, Taylor & Fancis, London.

Cracknell A P, Xue Y, 1996. Dynamic aspects study of surface temperature from remotely-sensed data using advanced thermal inertial model. *International Journal of Remote Sensing*, 17(13): 2 517-2 532.

Crank J, Nicolson P, 1947. A practical method for numerical evaluation of solutions of partial differential equations of the heat conduction types. *Proceeding of Cambridge Philosophy Society*, 43: 50-67.

Cuenca H R, Brouwer J, Chanzy A, et al., 1997. Soil measurements during HAPEX-Sahel intensive observation period. *Journal of hydrology*, 188: 224-266.

Curran P J, 1985. *Principles of Remote Sensing*, Longman Ltd., London, UK.

Danin A, 1983. *Desert Vegetation of Israel and Sinai*, Cana Publishing House, Jerusalem, Israel.

Danin A, 1987. Impact of man on biological components of desert ecosystems in Israel. *XIV International Botanical Congress*, Berlin (West), Germany.

Danin A, 1991. Plant adaptation in desert dunes. *Journal of Arid Environment*, 21: 193-212.

Danin A, Bar-Or Y, Dor I, et al., 1989. The role of cyanobacteria in stableilization of sand dunes in southern Israel. *Ecological Mediterranean*, 15: 55-64.

Davies J A, Idso S B, 1979. Estimating the surface radiation balance and its components. *Modification of the Aerial Environment of Plants*, American Society of Agricultural Engineers, St. Joseph, Michigan, USA.

Day P R, Luthin J N, 1956. A numerical solution of the differential equation of flow for a vertical drainage problem. *Proceeding of Soil Science Society of America*, 20: 443-447.

De Vries D A, 1963. Thermal properties of soils. *Physics of Plant Environment*, North-Holland Publishing Company, Amsterdam, The Netherlands.

De Lira G R, Batchily K, Huete A R, 1994. Optical and seasonal variations along the US-Mexico border: an analysis with Landsat TM imagery. IEEE, 1 044-1 045.

Diak G R, Whipple M S, 1993. Improvements to models and methods for evaluating the land surface energy balance and 'effective' roughness using radiasonde reports and satellite-measured 'skin' temperature data. *Agricultural and Forest Meteorology*, 63: 189-218.

Dolman A J, 1993. A multiple-source land surface energy balance model for use in general circulation models. *Agricultural and Forest Meteorology*, 65: 21-45.

Dolman A J, Blyth E M, 1997. Patch scale aggregation of heterogeneous land surface cover for mesoscale meteorological models. *Journal of Hydrology*, 190: 252-268.

Dor I, Furer O, Adin A, et al., 1993. Turbidity related to surface temperature in oxidation ponds: studies toward development of a remote sensing method. *Water Science and Technology*, 27: 37-44.

Drury S A, 1990. *A Guide to Remote Sensing: Interpreting images of the earth*, Oxford University Press, Oxford, UK.

Dunbier R, 1968. *The Sonoran Desert: its geography, economy and people*, University of Arizona Press, Tucson, Arizona, USA.

Elrick D E, Bowman D H, 1964. Note on an improved apparatus for soil moisture flow measurements, *Proceeding of Soil Science Society of America*, 28: 450-452.

Evenari M, Shana L, Tadmor N, 1971. *The Negev - the challenge of a desert*, Harvard University Press, Cambridge, Massachusetts, USA.

Fania C, Zipora G, 1997. *Soil Temperature Regime in Israel*, Israel Meteorological Service, Bet Dagon, Israel.

Farrand W H, Singer R B, Merenyi E, 1994. Retrieval of apparent surface reflectance from AVIRIS data: a comparison of empirical line, radiative transfer and spectral mixture methods. *Remote Sensing of Environment*, 47:311-321.

Flanders H, Price J J, 1978. *Calculus with Analytic Geometry*, Academic Press, New York, USA.

França G B, and Cracknell A P, 1994. Retrieval of land and sea surface temperature using NOAA-11 AVHRR data in northeastern Brazil. *International Journal of Remote Sensing*, 15(8): 1 695-1 712.

Gao B C, Heidebecht K B, Goetz, A F H, 1998. Derivation of scaled surface reflectance from AVIRIS data. *Remote Sensing of Environment*, 44:165-178.

Gardner C M K, Bell J D, Cooper J D, et al., 1991. Soil water content. *Soil Analysis*, 1-74, Marcel Dekker Inc., New York, USA.

Gardner W H, 1965. Water content. *Methods of Soil Analysis*, 82-127, American Society of Agronomy, Madison, Wisconsin, USA.

Garratt J R, 1984. The measurement of evaporation by meteorological methods. *Agricultural Water Management*, 8: 99-117.

Gerson R, Amit R, Grossman S, 1985. *Dust Availability in Desert Terrain: a study in the deserts of Israel and the Sinai*, The Hebrew University of Jerusalem, Israel.

Ghildyal B P, Tripathi P, 1987. *Soil Physics*, John Wiley & Sons Ltd., New York, USA.

Givri J R, 1995. Assessing the relation between emissivity and vegetation with AVHRR. *International Journal of Remote Sensing*, 16: 2 971-2 988.

Grant R F, Izaurralde R C, Chanasyk D S, 1995. Soil temperature under different surface managements: testing a simulation model. *International Journal of Remote Sensing*, 73: 89-113.

Gupta R P, 1991. *Remote Sensing Geology*, Springer Verlag, Berlin Heidelberg.

Ham J M, Kluitenberg G J, 1993. Positional Variation in the soil energy balance beneath a row-crop canopy. *Agricultural and Forest Meteorology*, 63: 73-92.

Hanks R J, Ashcroft G L, 1980. *Applied Soil Physics: Soil water and temperature applications*, Springer-Verlag Berlin Heidelberg Ltd., New York, USA.

Harrison A R, Melia J, Bastida J, et al., 1996. Remote sensing of Mediterranean vegetation and surface lithology. *Mediterranean Desertification and Land Use*, John Wiley & Sons Ltd., West Sussex, England, UK.

Harrison B A, Jupp D L B, 1990. *Introduction to Remotely Sensed Data*, CSIRO, Australia.

Harrys R, 1987. *Satellite Remote Sensing: an introduction*, Routledge & Kegan Paul Ltd., London, UK.

Hausenbuiller R I, 1978. *Soil Science*, Wm C. Brown Company, Dubuque, Iowa, USA.

Heilman J L, McInnes K J, Savage M J, et al., 1995. Soil and canopy energy balance in west Texas vineyard. *Agricultural and Forest Meteorology*, 71: 99-114.

Hillel D, 1971. *Soil and Water - Physical principles and processes*, Academic Press, Now York.

Hillel D, 1977. *Computer Simulation of Soil-Water Dynamics: a compendium of recent work*, International Development Research Center, Ottawa, Canada.

Hillel D, 1980a. *Application of Soil Physics*, Academic Press, New York, USA.

Hillel D, 1980b. *Fundamentals of Soil Physics*, Academic Press, New York, USA.

Holbo H R, Luvall J C, 1989. Modeling surface temperature distributions in forest landscapes. *Remote Sensing of Environment*, 27(1): 11-24.

Hook S J, Gabell A R, Green A A, et al., 1992. A comparison of techniques for extracting emissivity information from thermal infrared data for geologic studies. *Remote Sensing of Environment*, 42:123-135.

Hord R M, 1986. *Remote Sensing: methods and applications*, John Wiley & Sons, Inc., New York, USA.

Humes K S, Kustas W P, Moran M S, et al., 1994. Variability of emissivity and surface temperature over a sparsely vegetated surface. *Water Resources Research*, 30(5): 1 299-1 310.

Hurtado E, Vidal A, Caselles V, 1996. Comparison of two atmospheric correction methods for Landsat TM thermal band. *International Journal of Remote Sensing*, 17: 237-247.

Idso S B, Jackson R D, Ehrler W L, et al., 1969. A method for determination of infrared emittances of leaves, *Ecology*, 50: 899-902.

Ignatov A M, Dergileva I L, 1995. Model approximation of angular dependence in AVHRR brightness temperature over a black surface. *International Journal of Remote Sensing*, 16: 3687-3693.

Kahle A B, 1980. Surface thermal properties. *Remote Sensing in Geology*, 257-273, John Wiley, New York.

Karnieli A, 1997. Development and implementation of spectral crust index over dune sands. *International Journal of Remote Sensing*, 18(6): 1 207-1 220.

Karnieli A, Shachack M, Tsoar H, et al., 1996. The effect of microphytes on the spectral reflectance of vegetation in semi-arid regions, *Remote Sensing of Environment*, 57:88-96.

Karnieli, A, Sarafis V, 1996. Reflectance spectrophotometry of cyanobacteria within soil crusts-a diagnostic tool. *International Journal of Remote Sensing*, 17(8): 1 609-1 615.

Karnieli A, Tsoar H, 1995. Spectral reflectance of biogenic crust developed on desert dune sand along the Israel-Egypt border. *International Journal of Remote Sensing*, 16: 369-374.

Kaufman Y J, 1989. The atmospheric effect on remote sensing and its correction. *Theory and Applications of Optical Remote Sensing*, John Wiley & Sons, New York.

Kerr Y H, Lagouarde J P, Imbernon J, 1992. Accurate land surface temperature retrieval from AVHRR data with use of an improved split window algorithm. *Remote Sensing of Environment*, 41: 197-209.

Kidron G J, 1999. Altitude dependent dew and fog in the Negev Desert, Israel. *Agricultural and Forest Meteorology*, 96: 1-8.

Kidron G J, Yair A, 1997. Rainfall-runoff relationship over encrusted dune surfaces, Nizzana, Western Negev, Israel. *Earth Surface Processes and Landforms*, 22: 1 169-1 184.

Kimball B A, Jackson R D, 1979. Soil heat flux. *Modification of the Aerial Environment of Plants*, American Society of Agricultural Engineers, St. Joseph, Michigan, USA.

Klute A, 1973. Soil water flow theory and its application in field situation. *Field Soil Water Regime*, Soil Science society of America, Madison Wisconsin, USA.

Koorevaar P, Menelik G, Dirksen C, 1983. *Elements of soil physics*, Elsevier Science Publishers, Amsterdam, Holand.

Kramer M, 1985. US-Mexican border: life on the line, *National Geography*, 167: 6.

Kustas W P, Jackson R D, Asrar G, 1989. Estimating surface energy balance components from remotely sensed data. *Theory and Applications of Optical Remote Sensing*, John Wiley & Sons, New York.

Kutilek M, Nielson D R, 1994. *Soil Hydrology*, Catena Verlag Ltd., Destedt, Germany.

Labed J, Stoll M P, 1991a. Angular variation of land surface spectral emissivity in the thermal infrared: laboratory investigations on bare soils. *International Journal of Remote Sensing*, 12(11): 2 299-2 310.

Labed J, Stoll M P, 1991b. Spatial variability of land surface emissivity in the thermal infrared band: spectral signature and effective surface temperature. *Remote Sensing of Environment*, 38(1): 1-17.

Lemon E R, Stewart D W, Shawcroft R W, et al., 1973. Experiments in predicting evaportranspiration by simulation with a soil-plant-atmosphere model. *Field Soil Water Regime*, Soil Science society of America, Madison Wisconsin, USA.

Lhomme J P, Monteny B, Amadou M, 1994. Estimating sensible heat flux from radiometric temperature over sparse millet. *Agricultural and Forestry Meteorology*, 68: 77-91.

Li Z L, Becker F, 1993, Feasibility of land surface temperature and emissivity determination from AVHRR data. *Remote Sensing of Environment*, 43(1): 67-85.

Lillesand T M, Kiefer R W, 1987. *Remote Sensing and Image Interpretation*, John Wiley & Sons, Ltd.

Markham B L, Barker J L, 1986. Landsat-MSS and TM post calibration dynamic ranges, atmospheric reflectance and at-satellite temperature. *EOSAT Landsat Technical Notes 1*, Earth Observation Satellite Company, Lanham, Maryland, USA.

Marshall T J, Holmes J W, 1979. *Soil Physics*, Cambridge University Press, Cambridge, UK.

McInnes K J, 1981. *Thermal conductivity of soils from dry-land wheat regions of Eastern Washington*. M.S. Thesis, Washington State University, Pullman, USA.

McInnes K J, Kanemasu E T, Kissel D E, et al., 1986. Predicting diurnal variations in water content along with temperature at the soil surface. *Agricultural and Forest Meteorology*, 38: 337-348.

McMillin L M, 1975. Estimation of sea surface temperatures from two infrared window measurements with different absorption. *Journal of Geophysical Research*, 36: 5 113-5 117.

Menenti M, 1984. *Physical Aspects and Determination of Evaporation in Deserts Applying Remote Sensing Techniques*, Institute of Land-Water Management and Resources (ICW), Wageningen, The Netherland.

Monteith J L, 1973. *Principles of Environmental Physics*, Edward Arnold Ltd., London, UK.

Moreshet S, Cohen I, Fuchs M, 1983. Response of mature "Shamouti" Orange trees to irrigation of different soil volumes at similar levels of available water. *Irrigation Science*, 3: 223-236.

Norman J M, Divakarla M, Goel N S, 1995. Algorithms for extracting information from remote thermal-IR observations of the Earth's surface. *Remote Sensing of Environment*, 51: 157-168.

Olioso A, 1995. Simulating the relationship between thermal emissivity and the Normalized Difference Vegetation Index. *International Journal of Remote Sensing*, 16: 3 211-3 216.

Olufayo A, Baldy C, Ruelle P, et al., 1993. Diurnal course of canopy temperature and leaf water potential of sorghum under a Mediterranean climate, *Agricultural and Forest Meteorology*, 64: 223-236.

Ottle C, Vidal-Madjar D, 1992. Estimation of land surface temperature with NOAA-9 Data, *Remote Sensing of Environment*, 40(1): 27-41.

Otterman J, 1974. Baring high-albedo soils by overgrazing: a hypothesized desertification mechanism. *Science*, 186: 531-533.

Otterman J, 1977. Anthropogenic impact on the surface at the Earth. *Climatic Change*, 1: 137-155.

Otterman J, 1981. Satellite and field studies of man's impact on the surface in arid regions. *Tellus*, 33: 68-77.

Otterman J, Fraser R S, 1976. Earth-atmosphere system and surface reflectivity in arid regions from

Landsat multi-spectral scanner measurements. *Remote Sensing of Environment*, 5: 247-266.

Otterman J, Robinove C J, 1982. Landsat monitoring of desert vegetation growth in 1972-1979 using a plant shadowing model. *Advance of Space Researches*, 2, 45-50.

Otterman J, Tucker C J, 1985. Satellite measurements of surface albedo in semi desert. *Journal of Climate and Applied Meteorology*, 24: 228-235.

Otterman J, Waisel Y, Rosenberg E, 1975. Western Negev and Sinai ecosystems: comparative study of vegetation, albedo, and temperatures. *Agro-Ecosystems*, 2: 47-59.

Ottlé C, Stoll M, 1993. Effect of atmospheric absorption and surface emissivity on the determination of land surface temperature from infrared satellite data. *International Journal of Remote Sensing*, 14: 2 025-2 037.

Pinker RT, Ferrare R A, Karnieli A, et al., 1997. Aerosol optical depths in a semiarid region. *Journal of Geophysical Research*, 102: 11 123-11 137.

Pinker R T, Karnieli A, 1995. Characteristic spectral reflectance of a semi-arid environment. *International Journal of Remote Sensing*, 16: 1 341-1 363.

Prata A J, 1993. Land surface temperature from the advanced very high resolution radiometer and the along-track scanning radiometer. 1. Theory. *Journal of Geographical Research*, 98: 16 689-16 702.

Prata A J, 1994. Land surface temperature from the advanced very high resolution radiometer and the along-track scanning radiometer. 2. Experimental results and validation of AVHRR algorithms. *Journal of Geographical Research*, 99: 13025-13058.

Prata A J, Caselles V, Coll C, et al., 1995. Thermal remote sensing of land surface temperature from satellites: current status and future prospects. *Remote Sensing Reviews*, 12: 175-224.

Price J C, 1983. Estimating surface temperature from satellite thermal infrared data - A simple formulation for the atmospheric effect. *Remote Sensing of Environment*, 13: 353-361.

Price J C, 1984. Land surface temperature measurements from the split window channels of the NOAA 7 Advanced Very High Resolution Radiometer. *Journal of Geophysical Research*, 89: 7 231-7 237.

Price J C, 1989. Quantitative aspects of remote sensing in the thermal infrared. *Theory and Applications of Optical Remote Sensing*, 578-603, John Wiley & Sons, New York.

Qin Z, Li W, Burgheimer J, et al., 2006a. Quantitative estimation of land cover structure in an arid environment across Israel-Egypt border using remote sensing data. *Journal of Arid Environment*. 66(2): 336-352.

Qin Z, Li W, Gao M, et al., 2006b. Estimation of land surface emissivity for the wavelength of Landsat TM6 and its application to Lingxian Region in north China Plain. SPIE 2006 Proceeding of Remote Sensing for Environmental Monitoring, GIS Applications, and Geology, 6366: 1-8.

Qin Z, Karnieli A, Berliner P, 2005. Ground temperature measurement and emissivity determination to understand the thermal anomaly and its significance on arid ecosystem development in the sand dunes across the Israel-Egypt border. *Journal of Arid Environment*, 60(1): 27-52.

Qin Z, Li W, Karnieli A, et al., 2003. Quantitative estimation of main land cover patterns in an arid environmental ecosystem across Israel-Egypt border using remote sensing data. *IEEE 2003 International Geosciences and Remote Sensing Symposium, III* Toulouse, France.

Qin Z, Karnieli A, Berliner P, 2002. Remote sensing analysis of the land surface temperature anomaly in the sand dune region across the Israel-Egypt border. *International Journal of Remote Sensing*, 23(19): 3 991-4 018.

Qin Z, Berliner P, Karnieli A, 2002. Numerical solution of a complete surface energy balance model for simulation of heat fluxes and surface temperature under bare soil environment. *Applied Mathematics and Computation*, 130(1):171-200.

Qin Z, Berliner P, Karnieli A, 2002. Micrometeorological modelling to understand the thermal anomaly in the sand dunes across the Israel-Egypt border. *Journal of Arid Environment*, 51(2):281-318.

Qin Z, Karnieli A, Berliner P, 2001. Thermal variation in the Israel-Sinai (Egypt) peninsula region. *International Journal of Remote Sensing*, 22(6): 915-919.

Qin Z, Dall'Olmo G, Karnieli A, et al., 2001. Derivation of split window algorithm and its sensitivity analysis for retrieving land surface temperature from NOAA-advanced very high resolution radiometer data. *Journal of Geophysical Research*, 106(D19), 22 655-22 670.

Qin Z, Karnieli A, Berliner P, 2001. A mono-window algorithm for retrieving land surface temperature from Landsat TM data and its application to the Israel-Egypt border region. *International Journal of Remote Sensing*, 22 (18): 3 719-3 746.

Qin Z, Karnieli A, 1999. Progress in the remote sensing of land surface temperature and ground emissivity using NOAA-AVHRR. *International Journal of Remote Sensing*, 20(12): 2 367-2 393.

Rendell H M, Yair A, Tsoar H, 1993. Thermoluminescence dating of periods of sand movement and linear dune formation in the northern Negev, Israel. *The Dynamics and environmental context of aeolian sedimentary system*, 72: 69-74.

Reutter H, Olesen F S, Fischer H, 1994. Distribution of the brightness temperature of land surfaces determined from AVHRR data. *International Journal of Remote Sensing*, 15: 95-104.

Richards L A, 1965. Physical condition of water in soil. *Methods of Soil Analysis*, American Society of Agronomy, Madison, Wisconsin, USA.

Richason B F, 1978. *Introduction to Remote Sensing of the Environment*, Kendall/Hunt Publishing Company, Dubuque, Iowa, USA.

Rose C W, 1969. *Agricultural Physics*, Pergamon Press Ltd., Oxford, Britain.

Sabins F F, 1980. Interpretation of thermal infrared images. *Remote Sensing in Geology*, 275-295, John Wiley Inc., New York, USA.

Sabins F F, 1987. *Remote Sensing Principles and Interpretation*, 2nd ed. Freeman, W H Inc., San Francisco.

Saraf A K, Prakash A, Sengupta S, et al., 1990. Landat-TM data for estimating ground temperature and depth of subsurface coal fire in the Jharia coalfield, India. *International Journal of Remote Sensing*, 16: 2 111-2 124.

Sceicz G, Endrodi G, Tachman S, 1969. Aerodynamic and surface factors in evaporation. *Water Resources Research*, 5: 380-394.

Schneider K, Mauser W, 1996. Processing and accuracy of Landsat Thematic Mapper data for lake surface temperature measurement. *International Journal of Remote Sensing*, 17(11): 2 027-2 041.

Schmugge T, Humes K, 1995. ASTER observations for the monitoring of land surface fluxes. *Journal of Japanese Remote Sensing Society*, 15: 83-89.

Schmugge T J, Becker F, Li Z L, 1991. Spectral emissivity variations observed in airborne surface temperature measurements. *Remote Sensing of Environment*, 35: 95-104.

Schott J R, Volchok W J, 1985. Thematic Mapper thermal infrared calibration. *Photogrammetric Engineering and Remote Sensing*, 51: 1 351-1 357.

Schultz P A, Halpert M S, 1995. Global analysis of the relationships among a vegetation index, precipitation and land surface temperature. *International Journal of Remote Sensing*, 16(15): 2 755-2 777.

Seguin B, 1996. The use of AVHRR-derived land surface temperature estimates for agricultural monitoring. *Advances in the Use of NOAA-AVHRR Data for Land Applications*, Kluwer Academic Publishers.

Seguin B, Courault D, Guerif M, 1994. Surface temperature and Evapotranspiration: application of local scale methods to regional scales using satellite data. *Remote Sensing of Environment*, 49(3): 287-295.

Sellers W D, 1965. *Physical Climatology*, The University of Chicago Press, Chicago.

Singh S M, 1988. Brightness temperature algorithms for Landsat Thematic Mapper Data. *Remote Sensing of Environment*, 24: 509-512.

Slater P N, 1980. *Remote Sensing, Optics and Optical Systems*. Addison-Wesley Company, Reading, Massachusetts, USA.

Smith G D, 1978. *Numerical Solution of Partial differential Equations: Finite difference methods*, Oxford University Press, Oxford, Britain.

Sobrino J A, Caselles V, 1991. A methodology for obtaining the crop temperature from NOAA-9 AVHRR data. *International Journal of Remote Sensing*, 12(12): 2 461-2 475.

Sobrino J A, Coll C, Caselles V, 1991. Atmospheric correction for land surface temperature using NOAA-11 AVHRR channels 4 and 5. *Remote Sensing of Environment*, 38(1): 19-34.

Sobrino J A, Li Z L, Stoll M P, et al., 1996. Multi-channel and multi-angle algorithms for estimating sea and land surface temperature with ATSR data. *International Journal of Remote Sensing*, 17(11): 2 089-2 114.

Sospedra F, Caselles V, Valor E, 1998, Effective wavenumber for thermal infrared bands application to Landsat TM. *International Journal of Remote Sensing*, 19:2 105-2 117.

Sullivan J T, 1995. A simple accurate method to compute brightness temperature and/or radiance for the thermal infrared channels of the Advanced Very High Resolution Radiometer. *International Journal of Remote Sensing*, 16(4): 773-777.

Sugita M, Brutsert W, 1993. Comparison of land surface temperature derived from satellite observation with ground truth during FIFE. *International Journal of Remote Sensing*, 14(9): 1 659-1 676.

Sutherland R A, 1986. Broadband and spectral (2-18μm) emissivity of some natural soils and vegetation. *Journal of Atmospheric and Oceanic Technology*, 3: 199-202.

Swain P H, Davis S M, 1978. *Remote Sensing: the quantitative approach*, McGraw-hill Inc., West Lafayette, Indiana, USA.

Swinbank W C, 1963. Long-wave radiation from clear skies. *Quarterly Journal of Royal Meteorological Society*, 8: 339-348.

Szekielda K H, 1988. *Satellite Monitoring of the Earth*, John Wiley & Sons Ltd., New York, USA

Takashima T, Masuda K, 1987, Emissivities of quartz and Sahara dust powers in the infrared region (1-17μm). *Remote Sensing of Environment*, 23: 51-63.

Thome K, Markham B, Barker J, et al., 1997. Radiometric calibration of Landsat. *Photogrammetric Engineering and Remote Sensing*, 63(7): 853-858.

Tielböger K, 1997. The vegetation of linear desert dunes in the northwestern Negev, Israel. *Flora*, 192: 261-278.

Tilton J C, 1991. *Multisource Data Interpretation in Remote Sensing*, NASA, USA.

Troeh F R, Jabro J D, Kirkham D, 1982. Gaseous diffusion equations for porous materials. *Geoderma*, 27: 239-253.

Tsoar H, 1990. Grain-size characteristics of wind ripples on a desert seif dune. *Geographic Research Forum*, 10: 37-50.

Tsoar H, 1995. Desertification on in northern Sinai in eighteenth century. *Climate Change*, 29: 429-438.

Tsoar H, Goldsmith V, Schoenhaus S, et al., 1995. Reversed desertification on sand dunes along the Sinai/Negev border. *Desert Aeolian Process*, Chapman and Hall, New York, USA.

Tsoar H, Karnieli A, 1996. What determines the spectral reflectance of the Negev-Sinai sand dunes. *International Journal of Remote Sensing*, 17(3): 3 513-3 525.

Tsoar H, Møller J T, 1986. The role of vegetation in the formation of linear sand dunes. *Aeolian Geomorphology*, Allen & Unwin Inc., Boston.

Van Bavel C H M, Hillel D I, 1976. Calculating potential and actual evaporation from a bare soil surface by simulation of concurrent flow of water and heat. *Agricultural and Forest Meteorology*, 17: 453-476.

Verma S B, Barfield B J, 1979. Aerial and crop resistance affecting energy transport. *Modification of the Aerial Environment of Plants*, American Society of Agricultural Engineers, St. Joseph, Michigan, USA.

Vidal A, 1991. Atmospheric and emissivity correction of land surface temperature measured from satellite ground measurements or satellite data. *International Journal of Remote Sensing*, 12: 2 449-2 460.

Vogt J V, 1996. Land surface temperature retrieval from NOAA-AVHRR data. *Advances in the Use of NOAA-AVHRR Data for Land Applications*, Kluwer Academic Publishers.

Waisel Y, 1986. Interactions among plants, man and climate: historical evidence from Israel. *Proceedings of the Royal Society of Edinburgh*, 89B: 255-264.

Wan Z, Dozier J, 1996. A generalized split-windows algorithm for retrieving land surface temperature from space. *IEEE Transactions on Geoscience and Remote Sensing*, 34(4): 892-905.

Warren A, Harrison C M, 1984. People and the ecosystem: Biogeography as a study of ecology and culture. *Geoforum*, 15: 365-381.

Wooster M J, Richards T S, Kidwell K, 1995. NOAA-11 AVHRR/2 - Thermal channel calibration update. *International Journal of Remote Sensing*, 16: 359-363.

Wukelic G E, Gibbons D E, Martucci L M, et al., 1989. Radiometric calibration of Landsat Thermatic Mapper Thermal Band. *Remote Sensing of Environment*, 28: 339-347.

Yair A, 1990. Runoff generation in a sand area - the Nizzana Sands, Western Negev, Israel. *Earth Surface Processes and Landforms*, 15, 597-609.

Yair A, Lavee H, Greitser N, 1997. Spatial and temporal variability of water percolation and movement in a system of longitudinal dunes, western Negev, Israel. *Hydrological Processes*, 11: 43-58.

Yakirevich A, Berliner P, Sorek S, 1997. A model for numerical simulating of evaporation from bare saline soil. *Water Resources Research*, 33: 1 021-1 033.

Yu Y, Barton I J, 1994. A non-regression-coefficients methods of sea surface temperature from space, *International Journal of Remote Sensing*, 15(6): 1 189-1 206.

Zemel Z, Lomas J, 1977. *Soil Temperature Regime in Israel as a Basis for Agricultural Planning and Activity*, Israel Meteorological Service, Bet Dagon, Israel.

Zhao W G, Berliner P R, Zangvil A, 1996. Heat storage items in evapotranspiration estimation. *Evapotranspiration and Irrigation Scheduling: proceedings of the international conference*, American Society of Agricultural Engineers.

Appendix: Computer program for the surface energy balance model

```
'Program for simulation of surface temperature
'through surface energy balance equation:
'simultaneously solve Ts and Ws from differential
'equations about soil heat transfer and water
movement

CLS : CLEAR
   nt = 60 'seconds in each time interval
   n = 24 ×60×2.5 + 1'number of time intervals
   m = 20 'number of the soil layers

'define arrays for soil profile parameters
   DIM ts0(m), ts1(m) 'soil temperature, K
   DIM vw0(m), vw1(m) 'soil water content
   DIM hs0(m), hs1(m) 'soil vapor humidity
   DIM ev(m) 'saturated vapor at soil layers
   DIM dv(m) 'vapor conductivity
   DIM ch(m) 'ratio of water content and
   humidity
   DIM sl(m) 'slope of temp and vapor pressure
   DIM kc(m) 'hydraulic conductivity
   DIM ks(m) 'soil thermal conductivity
   DIM cs(m) 'soil heat capacity
   DIM vq(m) 'quartz fraction
   DIM vm(m) 'clay mineral fraction
   DIM vo(m) 'soil organic fration
   DIM dz(m) 'thickness of soil layer

'define arrays for Crank-Nicolson method
'arrays for soil water movement equation
DIM ma(m), mb(m), mc(m), mg(m)
DIM md(m), me(m), mf(m)
'arrays for soil heat transfer equation
DIM aa(m), bb(m), cc(m), gg(m)
DIM dd(m), ee(m), ff(m)
'arrays for temperory use
DIM ma1(m), mb1(m), md1(m), me1(m),
mg1(m)
DIM aa1(m), bb1(m), dd1(m), ee1(m),
gg1(m)
DIM pn(m)

fin$ = "c:\qink\simulat\smniz.prn" 'file for input
fout$ = "c:\qink\simulat\smout.txt" 'file for output

'set the required meteorological constants
   stef = 5.67E-08 'Stefan-Boltz. constant, w/(m2
   K4)
   z = 2'standard height, m
   z0 = .02 'roughness length, m
   zz0 = LOG(z / z0) 'constant ln(z/z0)
   kc = .4 'Karman constant
   ag = 9.8'acceleration of gravity, m/s2
   gr = 461.52'gas constant, J/(kg K)
   lhv = 2453000 'laten heat vaporization of
   water, J/kg
   tcw = .6'thermal conductivity of water, w/(m·K)
   tcq = 8.8 'thermal conductivity of quartz, w/(m·K)
```

```
    tcm = 2.9 'thermal conduct. of clay minerals, w/(m·K)
    tca = .025 'thermal conduct. of air, w/(m·K)
    tco = 2.5 'thermal conduct of organic materials
    pt0 = -5 'J/kg
    bb = 2.5: bm = bb + 4: uu = .05: vu = 1.5'constants

'input some meteorological parameters
    ems = .97 'surface emissivity
    dai = 1.2923 'air density at 0C, kg/m^3
    dmi = 2700 'soil mineral density, kg/m^3
    dor = 1300 'soil organic density, kg/m^3
    dwa = 998.2 'water density, kg/m^3
    shmi = 733 'soil mineral specific heat, J/(kg·K)
    shor = 1926 'soil organic specific sheat, J/(kg ·K)
    shwa = 4182 'water specific heat, J/(kg·K)
    shai = 1005 'air specific heat, J/(kg·K)
    pa = 101.325 'atmosperic pressure, kPa

'assume some parameters
    tk = 273.16 'absolute temperature, K
    ep = 1 'accuracy for approximation
    pi = 3.141592654#'circle constant
'thickness of soil layer m
    FOR i = 1 TO m
    dz(i) = .025
    NEXT i
    dz(1)=0.005'biogenic crust
    'volumetric fraction soil components, m^3/m^3
    FOR i = 0 TO m
    vq(i) = .45 'quartz
    vm(i) = .05 'clay minerals
    vo(i) = 0 'organic materials
    NEXT i
    vq(0)=.05:vm(0)=.45:vo(0)=0'biogenic crust

'initial soil temperature C
ts0(0) = 40: ts0(m) = 29
FOR i = 1 TO m - 1
ts0(i) = 40 - i * 12.2 / m
NEXT i
    FOR i = 0 TO m
    ts0(i) = ts0(i) + tk
    NEXT i

FOR i = 0 TO m
vw0(i) = 8 + i * 4.7
NEXT i

'initial soil relative humidity
    FOR i = 0 TO m
    ssw = (1 - vq(i) - vm(i) - vo(i)) * dwa
    pti = pt0 * (ssw / vw0(i)) ^ bb
'relative humidity of soil gas
    hs0(i) = EXP(pti / (gr * ts0(i)))
    PRINT i, "hs0="; hs0(i); "ssw="; ssw; "pti="; pti
'IF i MOD 10 = 0 THEN INPUT fr
NEXT i

GOSUB 2000 'compute initial soil parameters
PRINT "i vw0 hs0 kc cs ks ev"
FOR i = 0 TO m
PRINT i; vw0(i); hs0(i); kc(i); cs(i); ks(i); ev(i)
NEXT i
'PRINT "i dv(i) sl(i) ch(i)"
'FOR i = 0 TO m
'PRINT i; dv(i); sl(i); ch(i)
'NEXT i
INPUT fr
OPEN fout$ FOR OUTPUT AS #1

PRINT #1, "0 Time Rs Rn H G LE wsp ra fyh fym ris Ta Ts0 Ts10 Ts20 Ts30 Ts40 Ts50 ";
PRINT #1, "ea es ha hs vw0 vw10 vw20 vw30 vw40 vw50 kc0 kc10 kc20 kc30 kc40 kc50 ";
```

```
PRINT #1, "ks0 ks10 ks20 ks30 ks40 ks50"

'open file for data input
OPEN fin$ FOR INPUT AS #2

'start the simulation for the time interval [1,n]
   FOR sl = 1 TO n
   INPUT #2, tm, rs, alsand, alcrust, ta, reh, wv
   ta = ta + tk: als = alsand
   PRINT sl, tm, rs, ta, reh, wv
'INPUT fr
'compute air vapor pressure
   evt = 26.6904 – 6109.74 / ta – 9.16189E-03 * ta
   eva = .1 * EXP(evt)'at saturation
   ea = reh * eva 'at unsaturation

'Newton'smethod to solve surface temperature
   ts1(0) = ts0(0) - .02 'initial temperature for
   iteration
   60 GOSUB 120 'calculate the energy balance
   equation
   fts1 = fts
   IF ABS(fts1) < 8 THEN 110
   ts1(0) = ts1(0) + .001

   GOSUB 120 'calculate the energy balance
   equation
   ts1(0) = ts1(0) - .001 - fts1 * .001 / (fts - fts1)
   PRINT "fts1="; fts1; "fts2="; fts
'INPUT fr
GOTO 60 'iteration to solve true surface temperature

110 'output the simulation results in a file
PRINT sl; "Time="; tm; " Rs="; rs; " Rn="; rn; " H=";
hh;
PRINT " G = "; sg; " LE = "; le; ; le0; le1;
"Ts="; ts1(0) - tk;
PRINT "Ta="; ta - tk; "hs="; hs1(0); "ha="; reh;
PRINT "wsp="; wv; "als="; als; "zlm="; zlm; " ra=";
ra;
PRINT "ris="; ris; " fyh="; fyh; " fym="; fym;
PRINT "es="; es; "ea="; ea; "vw="; vw1(0);

'output the results into a file
IF (sl - 1) MOD 10 = 0 THEN
PRINT #1, tm; rs; rn; hh; sg; le; wv; ra; fyh;
fym; ris; ta - tk;
FOR i = 0 TO m
IF i MOD 2 = 0 THEN PRINT #1, ts1(i) - tk;
NEXT i
PRINT #1, ea; es; reh; hs1(0);
FOR i = 0 TO m
IF i MOD 2 = 0 THEN PRINT #1, vw1(i);
NEXT i
FOR i = 0 TO m
IF i MOD 2 = 0 THEN PRINT #1, kc(i);
NEXT i
FOR i = 0 TO m
IF i MOD 2 = 0 THEN PRINT #1, ks(i);
NEXT i
PRINT #1,
END IF

'set the results of time j+1 into j for next iteration
FOR i = 0 TO m
ts0(i) = ts1(i): hs0(i) = hs1(i)
NEXT i

GOSUB 2000 'calculate soil parameters for next
iteration
PRINT "ev="; ev(0); "dv="; dv(0); "sl="; sl(0);
"kc="; kc(0);
PRINT "ks="; ks(0); "cs="; cs(0); "ch = "; ch(0)
PRINT
NEXT sl
CLOSE #2
```

```
CLOSE #1
END


120 'calculate energy balance equation for Ts
'calculate air emissivity at air temperature
   ema = .92 * .00001 * ta * ta

'calculate net radiation
   var1 = ema * stef * ta ^ 4
   var2 = ems * stef * ts1(0) ^ 4
   rn = rs * (1 - als) - var2 + ems * var1
'PRINT "Rn="; rn

'calculate heat flux
   var1 = dai * shai * (ts1(0) - ta)
   fyh = 0: fym = 0: hh0 = 0: zlm0 = 0
800 var2 = zz0 - fym: var3 = zz0 - fyh
   uv = wv * kc / var2
   ra = var3 / (kc * uv)
   hh = var1 / ra
   zlm = dai * shai * ta * uv * uv * uv
   zlm = -kc * z * ag * hh / zlm
   'PRINT "hh0="; hh0; "hh="; hh; "hh0-hh="; hh0 - hh
   IF ABS(hh0 - hh) < 2 * ep THEN 810
   hh0 = hh: zlm0 = zlm
   IF zlm >= zz0 THEN
   fyh = -5 * zz0: fym = fyh
   ELSEIF zlm >= 0 AND zlm < zz0 THEN
   fyh = -5 * zlm: fym = fyh
   ELSE
   xx = (1 - 16 * zlm) ^ .25
   fym = 2 * LOG(.5 + .5 * xx * xx)
   fyh = 2 * LOG(.5 + .5 * xx) + .5 * fym - 2 * ATN(xx) + .5 * pi
   END IF
   'PRINT "fyh="; fyh; "ts="; ts1(0); "ta="; ta
   GOTO 800
810 zlm = zlm0: hh = hh0
'PRINT "ra="; ra; "zlm="; zlm; "H="; hh

'calculate laten heat flux LE through evaporation

'solve LE, soil water content and temperature
   evt = 26.6904 - 6109.74 / ts1(0) - 9.16189E-03 * ts1(0)
   ev1 = .1 * EXP(evt)
   xx = .622 * dai * lhv
   'PRINT "vw0(0)="; vw0(0); "xx="; xx; "ev1="; ev1
      hs1(0) = hs0(0) - .0001
220 GOSUB 180 'solve hs & Ts of soil profile
   GOSUB 160 'compute soil water content
   GOSUB 165 'compute change of soil water content
   GOSUB 230 'solve LE by vapor pressure difference equation
   fle1 = le0 - le1
   'PRINT "fle1="; fle1
   'INPUT fr
   IF ABS(fle1) < 1 THEN 210
   hs1(0) = hs1(0) + .00001
   GOSUB 180
   GOSUB 160
   GOSUB 165
   GOSUB 230
   fle2 = le0 - le1
   hs1(0) = hs1(0) - .00001 - fle1 * .00001 / (fle2 - fle1)
   'PRINT "fle2="; fle2; "fle1="; fle1; "le0="; le0; "le1="; le1;
   IF hs1(0) < 0 THEN
   PRINT "Wrong...hs1(0)="; hs1(0)
   INPUT fr
   END IF
```

```
  GOTO 220

210 le = le0 'le = .5 * (le0 + le1)
  'PRINT "le="; le

'calculate heat flux into the soil G
  sg = 0
  FOR i = 1 TO m
  tsb = .5 * (ts1(i - 1) + ts1(i))
  tsa = .5 * (ts0(i - 1) + ts0(i))
  cs = .5 * (cs(i - 1) + cs(i))
  sg = sg + cs * (tsb - tsa) * dz(i)
  NEXT i
  sg = sg / nt'to make the unit in s.g.m system.
'PRINT "sg="; sg

'calculate the balance euqation
  'PRINT "Rn="; rn; "H="; hh; "G="; sg;
  "LE="; le
  fts = rn - hh - sg - le
RETURN

230 'comput LE by vapor pressure difference
equation
  es = hs1(0) * ev1
  'PRINT "hs1(0)="; hs1(0); "ts1(0)="; ts1(0)
  'saturated water content of surface
  ssw = (1 - vq(0) - vm(0) - vo(0)) * dwa
  'evaporation resistance of surface
  ris = 100 * (.4125 * ssw / vw1(0)) ^ 1.4445
  'PRINT "ea="; ea; "ris="; ris; "mvw1=";
  vw1(0)
  'latent heat flux
  le0 = xx * (es - ea) / (pa * (ra + ris))
  'PRINT "le0="; le0
  RETURN

160 'soil water content in soil for time j+1, kg/m^3
  FOR i = 0 TO m
'saturated water content
  ssw = (1 - vq(i) - vm(i) - vo(i)) * dwa
'soil water potential
  pti = gr * ts1(i) * LOG(hs1(i))
  bbb = 1 / bb
  pta = pti / pt0
  ptb = pta ^ bbb
'soil water content
  vw1(i) = ssw / ptb
  'PRINT i; "mvw0="; vw0(i); "mvw1=";
  vw1(i);
  'PRINT "hs0="; hs0(i); "hs1="; hs1(i)
  NEXT i
RETURN

165 'calculate soil water change from time j to j+1
  le1 = 0
  FOR i = 1 TO m
  a1 = (vw0(i - 1) + vw0(i)) * .5
  a2 = (vw1(i - 1) + vw1(i)) * .5
  le1 = le1 + (a1 - a2) * dz(i)
  NEXT i
  'PRINT "water lose="; lee1;
  le1 = lhv * le1 / nt
  'PRINT "le1="; le1
RETURN

2000 'calculate soil parameters
  FOR i = 0 TO m
'soil water content at saturation, kg/m^3
  ssw = (1 - vq(i) - vm(i) - vo(i)) * dwa
'air fraction in the soil
  vai = (ssw - vw0(i)) / dwa
  'PRINT "vai="; vai
'apparent vapor diffusivity
  dva = (2.22 + .158 * (ts0(i) - tk)) * .00001
  dv(i) = dva * ((1 - uu) / (vai - uu)) ^ vu
```

```
    dv(i) = dv(i) * .622 * dai / pa
'satured vapor pressure
    evt = 26.6904 - 6109.74 / ts0(i) - 9.16189E-03 * ts0(i)
    ev(i) = .1 * EXP(evt)
'soil vapor pressure
    esi = ev(i) * hs0(i)
'slope of saturated vapor pressure vs temperature
    sl(i) = 5307 * ev(i) / (ts0(i) * ts0(i))
'soil capillary (hydraulic) conductivity
    kh = .002 * EXP(-4.26 * vm(i))
    kc(i) = kh * (vw0(i) / ssw) ^ bm
    'PRINT "kc="; kc(i)
'soil heat capacity
    cs(i) = dmi * shmi * (vq(i) + vm(i)) + shwa * vw0(i)
    cs(i) = cs(i) + dor * shor * vo(i) + dai * shai * vai
    'PRINT "cs="; cs(i)
'soil thermal conductivity
    vs = 1 - vai
    tcg = tca + .075 * esi / pa
    fw = vs / 2.453: fq = vs * .051 / 2.453
    fm = vs * .104 / 2.453: fo = vs * 1.298 / 2.453
    fa = vai
    'PRINT "vs="; vs; "tcg="; tcg; "fw="; fw; "fq="; fq
    'PRINT "fm="; fm; "fo="; fo; "fa="; fa
    var1 = fw * vw0(i) * tcw / dwa + fq * vq(i) * tcq
    var1 = var1 + fm * vm(i) * tcm + fo * vo(i) * tco
    var1 = var1 + fa * va(i) * tcg
    var2 = fw * vw0(i) / dwa + fq * vq(i) + fm * vm(i)
    var2 = var2 + fo * vo(i) + fa * vai
    ks(i) = var1 / var2
    'PRINT "ks="; ks(i)
NEXT i
'INPUT fr

'ratio of soil water content to relative humidity
    FOR i = 0 TO m
    IF i = m THEN
    ch(i) = ch(i - 1) + (ch(i - 1) - ch(i - 3)) / 2
    ELSE
    ch(i) = (vw0(i + 1) - vw0(i)) / (hs0(i + 1) - hs0(i))
    END IF
    NEXT i

RETURN

180 'Crank-Nicolson method for heat transfer equation
    'compute the coefficients of the expansion equations

    FOR i = 0 TO m
    ma1(i) = gr * kc(i) * LOG(hs0(i)) + hs0(i) * sl(i) * dv(i)
    mb1(i) = gr * kc(i) * ts0(i) / hs0(i) + ev(i) * dv(i)
    md1(i) = ks(i) + hs0(i) * sl(i) * lhv * dv(i)
    me1(i) = ev(i) * lhv * dv(i)
    NEXT i
    ts1(m) = ts0(m): hs1(m) = hs0(m)

    FOR i = 1 TO m - 1
    kaa = .5 * (ma1(i - 1) + ma1(i))
    kab = .5 * (ma1(i) + ma1(i + 1))
    kba = .5 * (mb1(i - 1) + mb1(i))
    kbb = .5 * (mb1(i) + mb1(i + 1))
    kda = .5 * (md1(i - 1) + md1(i))
    kdb = .5 * (md1(i) + md1(i + 1))
    kea = .5 * (me1(i - 1) + me1(i))
```

```
keb = .5 * (me1(i) + me1(i + 1))

zti = dz(i) * dz(i)

ma(i) = kaa: md(i) = kba: mg(i) = 0
mc(i) = kab: mf(i) = kbb
aa(i) = kda: dd(i) = kea: gg(i) = 0
cc(i) = kdb: ff(i) = keb

IF i = 1 THEN
ma(i) = 0: md(i) = 0
mg(i) = kaa * ts1(0) + kba * hs1(0)
aa(i) = 0: dd(i) = 0
gg(i) = kda * ts1(0) + kea * hs1(0)
END IF
IF i = m - 1 THEN
mc(i) = 0: mf(i) = 0
mg(i) = kab * ts0(m) + kbb * hs0(m)
cc(i) = 0: ff(i) = 0
gg(i) = kdb * ts0(m) + keb * hs0(m)
END IF

b1 = zti * (ch(i + 1) + ch(i)) / nt
b2 = zti * (cs(i + 1) + cs(i)) / nt

mb(i) = kaa + kab
me(i) = kba + kbb + b1

mg(i) = mg(i) + kab * (ts0(i + 1) - ts0(i))
mg(i) = mg(i) - kaa * (ts0(i) - ts0(i - 1))
mg(i) = mg(i) + kbb * (hs0(i + 1) - hs0(i))
mg(i) = mg(i) - kba * (hs0(i) - hs0(i - 1))
mg(i) = mg(i) + b1 * hs0(i)
mg(i) = mg(i) + 2 * dzi * ag * (kc(i + 1) - kc(i))

bb(i) = kda + kdb + b2
ee(i) = kea + keb
   gg(i) = gg(i) + kdb * (ts0(i + 1) - ts0(i))
   gg(i) = gg(i) - kda * (ts0(i) - ts0(i - 1))
   gg(i) = gg(i) + b2 * ts0(i)
   gg(i) = gg(i) + keb * (hs0(i + 1) - hs0(i))
   gg(i) = gg(i) - kea * (hs0(i) - hs0(i - 1))
   NEXT i

   'PRINT "ts1(0)="; ts1(0); "hs1(0)="; hs1(0)
'PRINT "i; ma(i); mb(i); mc(i); md(i); me(i); mf(i); mg(i)"
'FOR i = 1 TO m - 1
'PRINT i; ma(i); mb(i); mc(i); md(i); me(i); mf(i); mg(i)
'NEXT i
'INPUT fr
'PRINT "i; aa(i); bb(i); cc(i); dd(i); ee(i); ff(i); gg(i)"
'FOR i = 1 TO m - 1
'PRINT i; aa(i); bb(i); cc(i); dd(i); ee(i); ff(i); gg(i)
'NEXT i
'INPUT "waiting..."; fr

'Gauss's method to get the final solution: u(i,j)
   ma1(1) = mb(1): mb1(1) = mc(1): md1(1) = me(1)
   me1(1) = mf(1): mg1(1) = mg(1)
   aa1(1) = bb(1): bb1(1) = cc(1): dd1(1) = ee(1)
   ee1(1) = ff(1): gg1(1) = gg(1)

   FOR i = 2 TO m - 1

   b1 = mb(i) - mb1(i - 1) * ma(i) / ma1(i - 1)
   d1 = md(i) - md1(i - 1) * ma(i) / ma1(i - 1)
   e1 = me(i) - me1(i - 1) * ma(i) / ma1(i - 1)
   g1 = mg(i) + mg1(i - 1) * ma(i) / ma1(i - 1)
```

```
'PRINT i; "b1="; b1; " d1="; d1;
'PRINT " e1="; e1; " g1="; g1

   bb1 = bb(i) - bb1(i - 1) * aa(i) / aa1(i - 1)
   dd1 = dd(i) - dd1(i - 1) * aa(i) / aa1(i - 1)
   ee1 = ee(i) - ee1(i - 1) * aa(i) / aa1(i - 1)
   gg1 = gg(i) + gg1(i - 1) * aa(i) / aa1(i - 1)
'PRINT "bb1="; bb1; " dd1="; dd1;
'PRINT " ee1="; ee1; " gg1="; gg1

   ma1(i) = b1 - bb1 * d1 / dd1
   mb1(i) = mc(i) - cc(i) * d1 / dd1
   md1(i) = e1 - ee1 * d1 / dd1
   me1(i) = mf(i) - ff(i) * d1 / dd1
   mg1(i) = g1 - gg1 * d1 / dd1
'PRINT "a1="; ma1(i); " b1="; mb1(i);
'PRINT " d1="; md1(i); " g1="; mg1(i)

   b1 = bb(i) - mb1(i - 1) * aa(i) / ma1(i - 1)
   d1 = dd(i) - md1(i - 1) * aa(i) / ma1(i - 1)
   e1 = ee(i) - me1(i - 1) * aa(i) / ma1(i - 1)
   g1 = gg(i) + mg1(i - 1) * aa(i) / ma1(i - 1)
'PRINT "b1="; b1; " d1="; d1;
'PRINT " e1="; e1; " g1="; g1

   bb1 = mb(i) - bb1(i - 1) * ma(i) / aa1(i - 1)
   dd1 = md(i) - dd1(i - 1) * ma(i) / aa1(i - 1)
   ee1 = me(i) - ee1(i - 1) * ma(i) / aa1(i - 1)
   gg1 = mg(i) + gg1(i - 1) * ma(i) / aa1(i - 1)
'PRINT "bb1="; bb1; " dd1="; dd1;
'PRINT " ee1="; ee1; " gg1="; gg1

   aa1(i) = b1 - bb1 * d1 / dd1
   bb1(i) = cc(i) - mc(i) * d1 / dd1
   dd1(i) = e1 - ee1 * d1 / dd1
   ee1(i) = ff(i) - mf(i) * d1 / dd1
   gg1(i) = g1 - gg1 * d1 / dd1
'PRINT "aa1="; aa1(i); " bb1="; bb1(i);" dd1=";dd1(i);
'PRINT " ee1="; ee1(i); " gg1="; gg1(i)
   NEXT i


'PRINT "i; ma1(i); mb1(i); md1(i); me1(i); mg1(i)"
'PRINT "i; aa1(i); bb1(i); dd1(i); ee1(i); gg1(i)"
'FOR i = 1 TO m - 1
'PRINT i; ma1(i); mb1(i); md1(i); me1(i); mg1(i)
'PRINT i; aa1(i); bb1(i); dd1(i); ee1(i); gg1(i)
'NEXT i
'INPUT "waiting"; fr

   b1 = mg1(m - 1) * aa1(m - 1)
   b1 = b1 - gg1(m - 1) * ma1(m - 1)
   b2 = md1(m - 1) * aa1(m - 1)
   b2 = b2 - dd1(m - 1) * ma1(m - 1)
   'PRINT "@@@b1="; b1; "b2="; b2
   hs1(m - 1) = b1 / b2
   b1 = mg1(m - 1) * dd1(m - 1)
   b1 = b1 - gg1(m - 1) * md1(m - 1)
   b2 = ma1(m - 1) * dd1(m - 1)
   b2 = b2 - aa1(m - 1) * md1(m - 1)
   'PRINT "###b1="; b1; "b2="; b2
   ts1(m - 1) = b1 / b2

   FOR i = m - 2 TO 1 STEP -1
   g1 = mg1(i) + mb1(i) * ts1(i + 1)
   g1=g1 + me1(i) * hs1(i + 1)
   gg1 = gg1(i) + bb1(i) * ts1(i + 1)
   gg1=gg1 + ee1(i) * hs1(i + 1)
   b1 = g1 * aa1(i) - gg1 * ma1(i)
   b2 = md1(i) * aa1(i) - dd1(i) * ma1(i)
   hs1(i) = b1 / b2
   b1 = g1 * dd1(i) - gg1 * md1(i)
   b2 = ma1(i) * dd1(i) - aa1(i) * md1(i)
   ts1(i) = b1 / b2
```

```
NEXT i
ts1(m) = ts0(m): hs1(m) = hs0(m)

'FOR i = 1 TO m - 1
'tsa = ts1(i - 1): tsb = ts1(i): tsc = ts1(i + 1)
'hsa = hs1(i - 1): hsb = hs1(i): hsc = hs1(i + 1)
'IF i = 1 THEN tsa = 0: hsa = 0
'IF i = m - 1 THEN tsc = 0: hsc = 0
'eq1 = -ma(i) * tsa + mb(i) * tsb - mc(i) * tsc
'eq1 = eq1 - md(i) * hsa + me(i) * hsb - mf(i)
* hsc
'eq2 = -aa(i) * tsa + bb(i) * tsb - cc(i) * tsc
'eq2 = eq2 - dd(i) * hsa + ee(i) * hsb - ff(i) *
hsc
'PRINT i; "ts0="; ts0(i) - tk; "ts1="; ts1(i) -
tk;
'PRINT "hs0="; hs0(i); "hs1="; hs1(i)
'PRINT i; " eq1="; eq1; mg(i); "eq2="; eq2;
gg(i)
'IF i MOD 20 = 0 THEN INPUT fr
'NEXT i
RETURN
```